# Setting Priorities for Land Conservation

Committee on Scientific and Technical
Criteria for Federal Acquisition
Of Lands for Conservation

Board on Environmental Studies
And Toxicology

Commission on Life Sciences

NATIONAL RESEARCH COUNCIL
1993

**NATIONAL ACADEMY PRESS** 2101 Constitution Ave., N.W. Washington, D.C. 20418

NOTICE: The project that is the subject of this report was approved by the Governing Board of the National Research Council, whose members are drawn from the councils of the National Academy of Sciences, the National Academy of Engineering, and the Institute of Medicine. The members of the committee responsible for the report were chosen for their special competencies and with regard for appropriate balance.

This report has been reviewed by a group other than the authors according to procedures approved by a Report Review Committee consisting of members of the National Academy of Sciences, the National Academy of Engineering, and the Institute of Medicine.

The National Academy of Sciences is a private, non-profit, self-perpetuating society of distinguished scholars engaged in scientific and engineering research, dedicated to the furtherance of science and technology and to their use for the general welfare. Upon the authority of the charter granted to it by the Congress in 1863, the Academy has a mandate that requires it to advise the federal government on scientific and technical matters. Dr. Frank Press is president of the National Academy of Sciences.

The National Academy of Engineering was established in 1964, under the charter of the National Academy of Sciences, as a parallel organization of outstanding engineers. It is autonomous in its administration and in the selection of its members, sharing with the National Academy of Sciences the responsibility for advising the federal government. The National Academy of Engineering also sponsors engineering programs aimed at meeting national needs, encourages education and research, and recognizes the superior achievements of engineers. Dr. Robert M. White is president of the National Academy of Engineering.

The Institute of Medicine was established in 1970 by the National Academy of Sciences to secure the services of eminent members of appropriate professions in the examination of policy matters pertaining to the health of the public. The Institute acts under the responsibility given to the National Academy of Sciences by its congressional charter to be an adviser to the federal government and, upon its own initiative, to identify issues of medical care, research, and education. Dr. Kenneth I. Shine is president of the Institute of Medicine.

The National Research Council was organized by the National Academy of Sciences in 1916 to associate the broad community of science and technology with the Academy's purposes of furthering knowledge and advising the federal government. Functioning in accordance with general policies determined by the Academy, the Council has become the principal operating agency of both the National Academy of Sciences and the National Academy of Engineering in providing services to the government, the public, and the scientific and engineering communities. The Council is administered jointly by both Academies and the Institute of Medicine. Dr. Frank Press and Dr. Robert M. White are chairman and vice chairman, respectively, of the National Research Council.

The project was supported by Department of the Interior cooperative agreement no. 0660-9-8001.

**Library of Congress Catalog Card No. 92-62644**
**International Standard Book No. 0-309-04836-2**

B-057

Copyright 1993 by the National Academy of Sciences. All rights reserved.

Printed in the United States of America

# Committee on Scientific and Technical Criteria for Federal Acquisition of Lands For Conservation

**WILLIAM H. RODGERS, JR.** *(Chairman)*, University of Washington, Seattle, Wash.
**MICHAEL J. BEAN**, Environmental Defense Fund, Washington, D.C.
**HARRIET BURGESS**, American Land Conservancy, San Francisco, Calif.
**SALLY K. FAIRFAX**, University of California, Berkeley, Calif.
**CHARLES C. GEISLER**, Cornell University, Ithaca, N.Y.
**PERRY R. HAGENSTEIN**, Resource Issues, Inc., Wayland, Mass.
**LAWRENCE D. HARRIS**, University of Florida, Gainesville, Fla.
**ROBERT G. HEALY**, Duke University, Durham, N. Car.
**THOMAS E. LOVEJOY**, The Smithsonian Institution, Washington, D.C.
**JOHN P. MCMAHON**, Weyerhaeuser Company, Tacoma, Wash.
**DEBRA J. SALAZAR**, Western Washington University, Bellingham, Wash.
**WILLIAM W. SHAW**, University of Arizona, Tucson, Ariz.
**NANCY L. STANTON**, University of Wyoming, Laramie, Wyo.
**MONICA G. TURNER**, Oak Ridge National Laboratory, Oak Ridge, Tenn.
**CATHERINE VANDEMOER**, Council of Energy Resource Tribes, Denver, Colo.

*Staff*

**DAVID POLICANSKY**, Program Director
**LEE R. PAULSON**, Project Director (since July 1992)
**SYLVIA S. TOGNETTI**, Project Director (until July 1992)
**ANNE M. SPRAGUE**, Information Specialist
**HOLLY WELLS**, Senior Project Assistant

*Sponsor*: U.S. Department of the Interior

# Board on Environmental Studies and Toxicology

**PAUL G. RISSER** *(Chair)*, Miami University, Oxford, Ohio
**FREDERICK R. ANDERSON**, Cadwalader, Wickersham & Taft, Washington, D.C.
**JOHN C. BAILAR, III**, McGill University School of Medicine, Montreal, Quebec, Canada
**GARRY D. BREWER**, University of Michigan, Ann Arbor, Mich.
**JOHN CAIRNS, JR.**, Virginia Polytechnic Institute and State University, Blacksburg, Va.
**EDWIN H. CLARK**, Department of Natural Resources and Environmental Control, State of Delaware, Dover, Del.
**JOHN L. EMMERSON**, Lilly Research Laboratories, Greenfield, Ind.
**ROBERT C. FORNEY**, Unionville, Pa.
**ALFRED G. KNUDSON**, Fox Chase Cancer Center, Philadelphia, Pa.
**KAI N. LEE**, Williams College, Williamstown, Mass.
**GENE E. LIKENS**, The New York Botanical Garden, Millbrook, N.Y.
**JANE LUBCHENCO**, Oregon State University, Corvallis, Ore.
**DONALD R. MATTISON**, University of Pittsburgh, Pittsburgh, Pa.
**HAROLD A. MOONEY**, Stanford University, Stanford, Calif.
**GORDON ORIANS**, University of Washington, Seattle, Wash.
**FRANK PARKER**, Vanderbilt University, Nashville, Tenn.
**GEOFFREY PLACE**, Hilton Head, S. Car.
**MARGARET M. SEMINARIO**, AFL/CIO, Washington, D.C.
**I. GLENN SIPES**, University of Arizona, Tucson, Ariz.
**BAILUS WALKER, JR.**, University of Oklahoma, Oklahoma City, Okla.
**WALTER J. WEBER, JR.**, University of Michigan, Ann Arbor, Mich.

*Staff*

**JAMES J. REISA,** Director
**DAVID J. POLICANSKY,** Associate Director and Program Director for Natural Resources and Applied Ecology
**RICHARD D. THOMAS,** Associate Director and Program Director for Human Toxicology and Risk Assessment
**LEE R. PAULSON,** Program Director for Information Systems and Statistics
**RAYMOND A. WASSEL,** Program Director for Environmental Sciences and Engineering

# Commission on Life Sciences

BRUCE M. ALBERTS *(Chairman)*, University of California, San Francisco, Calif.
BRUCE N. AMES, University of California, Berkeley, Calif.
J. MICHAEL BISHOP, Hooper Research Foundation, University of California Medical Center, San Francisco, Calif.
DAVID BOTSTEIN, Stanford University School of Medicine, Stanford, Calif.
MICHAEL T. CLEGG, University of California, Riverside, Calif.
GLENN A. CROSBY, Washington State University, Pullman, Wash.
LEROY E. HOOD, University of Washington, Seattle, Wash.
MARIAN E. KOSHLAND, University of California, Berkeley, Calif.
RICHARD E. LENSKI, University of Oxford, Oxford, England
STEVEN P. PAKES, Southwestern Medical School at Dallas, Tex.
EMIL A. PFITZER, Hoffmann-La Roche Inc., Nutley, N.J.
MALCOLM C. PIKE, University of Southern California School of Medicine, Los Angeles, Calif.
THOMAS D. POLLARD, Johns Hopkins Medical School, Baltimore, Md.
PAUL G. RISSER, Miami University, Oxford, Ohio
JOHNATHAN M. SAMET, University of New Mexico School of Medicine, Albuquerque, N. Mex.
HAROLD M. SCHMECK, JR., Armonk, N.Y.
CARLA J. SHATZ, University of California, Berkeley, Calif.
SUSAN S. TAYLOR, University of California at San Diego, La Jolla, Calif.
P. ROY VAGELOS, Merck and Company, Inc., Rahway, N.J.
TORSTEN N. WIESEL, Rockefeller University, New York, N.Y.

*Staff*

ALVIN G. LAZEN, Acting Executive Director

# Other Recent Reports of the Board on Environmental Studies and Toxicology

Issues in Risk Assessment (1993)
Protecting Visibility in National Parks and Wilderness Areas (1993)
Hazardous Materials on the Public Lands (1992)
Dolphins and the Tuna Industry (1992)
Science and the National Parks (1992)
Biologic Markers in Immunotoxicology (1992)
Environmental Neurotoxicology (1992)
Animals as Sentinels of Environmental Health Hazards (1991)
Assessment of the U.S. Outer Continental Shelf Environmental Studies
    Program, Volumes I-IV (1991-1993)
Human Exposure Assessment for Airborne Pollutants (1991)
Monitoring Human Tissues for Toxic Substances (1991)
Rethinking the Ozone Problem in Urban and Regional Air Pollution
    (1991)
Decline of the Sea Turtles (1990)
Tracking Toxic Substances at Industrial Facilities (1990)
Biologic Markers in Pulmonary Toxicology (1989)
Biologic Markers in Reproductive Toxicology (1989)

*Copies of these reports may be ordered from*
*the National Academy Press*
*(800) 624-6242*

# Preface

The Committee on Scientific and Technical Criteria for Federal Acquisition of Lands for Conservation was formed under the auspices of the Board on Environmental Studies and Toxicology of the National Research Council's Commission on Life Sciences. Our charge was to review the criteria and procedures under which land is acquired for conservation by four of the federal land-management agencies—the National Park Service, the U.S. Forest Service, the U.S. Fish and Wildlife Service, and the Bureau of Land Management.

The subject is one of great complexity, and we saw disorder wherever we looked—in the definitions of criteria, procedures, and acquisition; in the histories, laws, and practices of the agencies; in the role of Congress; in the elaboration of the details of the acquisition transactions; in the difficulties of identifying who owns what in a world of partial and overlapping entitlements; in the methodologies used to describe, evaluate, and compare possible acquisitions; and in the mysteries of the social and natural sciences that stand in the way of firm predictions of whether acquisition X will achieve goal Y.

The field is so topsy-turvy that many of its most cherished assumptions must be set aside. Doubt now assails the scientific assumptions that conservation goals can be achieved indefinitely by property set-asides in the form of parks, preserves, and "arks." And questions arise on the policy front of whether land-acquisition goals can be realized by continued heavy reliance upon the crude tool of full-fee acquisition.

Land acquisition by government agencies raises a host of sociological, and inevitably political, issues of intense interest to numerous people—inholders, land-rights groups, acquisitions intermediaries, conservation organizations, state and local governments, and others. Indeed, the ultimate question of whether acquisition A should be given a higher priority than acquisition B is a political issue, because it boils down to a comparison of incommensurate values. The committee attempted to steer clear of this political thicket, and focus on description of the complex acquisition systems and on the technical and scientific aspects of the criteria. Congress makes political choices and exercises a strongly independent role in acquisitions, as the committee description shows. The chapter on the social effects of land acquisition illustrates, however, that topics of intense political controversy are not immune from illumination by scientific method.

The recommendations of this report can be described in large measure under the heading of "making connections" and improving integration. These include the recommendations to broaden the acquisition analysis from the single parcel to the landscape context; to link up piecemeal purchases to longer-term acquisition plans; to widen the scope of the acquisition techniques; to think in terms of corridors, connections, and linkages between properties; to identify holdings of other agencies and gaps in systems of protective ownership; and to sharpen the tools of acquisition to respond to emergent and opportunistic circumstance.

My personal appreciation is extended to the committee members who undertook the task with collegial enthusiasm and scientific objectivity. Their collective knowledge and experience cutting across many disciplines will be apparent to the readers of the report. The committee was guided and assisted in indispensable ways by the staff of the National Research Council. Sylvia Tognetti, project director until July 1992, was our bulldog, who did many of the basic research, writing, and coordination activities. David Policansky, program director, provided us guidance, perspective, and the judgmental interventions of the expert on science policy that he is. Lee Paulson, project director since July 1992, gave us the substantial editorial assistance that we needed and a welcome input of energy to carry us over the last hurdles to publication.

I would like to thank the individuals who made presentations to the committee and who provided us with statements and data. They include Henry Diamond, David Gibbons, Chip Collins, Matt Connolly, Jack Walter, Robert J. Smith, David Ford, John Heisenbuttel, Dean Bibles, Richard Moore, Michael Scott, Jerry Sutherland, Joseph Wrabek, Dale Crane, Bob Like, Chuck Williams, and Charles Jordan. We are grateful also the several anonymous reviewers of the report. They made especially strong contributions in their thankless task, and we benefited from their suggestions.

William H. Rodgers, Jr.<br>Chairman

# Contents

**EXECUTIVE SUMMARY**     *1*
    Considerations for Criteria, *3*
    Current Criteria, *5*
    Conclusions and Recommendations, *9*

**1 INTRODUCTION**     *15*
    Land Acquisition Agencies, *17*
    Guidelines for Criteria, *18*
    The Information Gap, *26*
    Valuation Challenges, *26*
    Report Organization, *27*

**2 PUBLIC LAND, PRIVATE LAND: AN OVERVIEW OF OWNERSHIP AND ITS MANAGEMENT CHALLENGES**     *29*
    Conservation: A Term of Many Meanings, *29*
    Public and Private Land Ownership, *39*
    Disincentives for Conservation, *47*
    The Role of Land Ownership in Conservation, *48*

**3 THE LAND ACQUISITION PROCESS**     *51*
    Sources of Funding, *51*
    Acquisition by Federal Agencies, *55*
    The Congress, *89*
    Landowners, *91*

Other Interested Parties, *92*
Rational Analysis and Politics in the Acquisition Process, *100*

**4 ASSESSING THE SOCIAL EFFECTS OF
FEDERAL LAND ACQUISITION** — *103*
Inholders and Federal Land Acquisition, *104*
Social Impact Assessment, *106*
SIA in Practice: A Bureau of Reclamation Case Study, *108*
Environmental Management and SIA, *110*
SIA and Conservation, *111*

**5 THE LAND ACQUISITION PROCESS
AND BIOLOGICAL PRESERVES:
A ROLE FOR NATURAL SCIENCES** — *113*
Fundamental Ecological Challenges, *113*
Other Biological Considerations, *127*
Enhancing the Ecological Effectiveness
  of the Acquisition Process, *128*
Conclusion, *135*

**6 NONPROFIT ORGANIZATIONS** — *139*
The National Fish and Wildlife Foundation, *141*
Ducks Unlimited, *151*
The Nature Conservancy, *153*

**7 TECHNIQUES AND TOOLS OF ACQUISITION** — *157*
Conservation Easements, *158*
Transferable Development Rights, *161*
Dedication, *162*
Regulation, *163*
Land Exchange, *166*
Land Acquisition Strategies and Transactions, *173*
Conclusion, *179*

**8 THE OFFICE OF MANAGEMENT AND BUDGET** — *183*
Adequacy of the LAPP Criteria, *184*
Reflection of Agency Missions and Authorities, *192*

**9 CONCLUSIONS AND RECOMMENDATIONS** *197*
Goals, *199*
Procedures, *204*

**REFERENCES** *213*
**APPENDIX A:** Presenters and Discussants *235*
**APPENDIX B:** Procedure for Compiling Federal
 Land Acquisition Priority List *237*
**APPENDIX C:** National Surveys Relevant to
 Public Land Use, Protection, and Purchase *243*
**APPENDIX D:** The Nature Conservancy:
 Acquisition Priorities and Preserve Selection and Design *247*
**GLOSSARY** *261*

# Executive Summary

*One million acres of the Grand Canyon, majestic as it is, isn't worth 100 acres of farmland in Iowa if you want to raise corn. They are not interchangeable.* [Former Secretary of the Interior Cecil Andrus.]

It is difficult to compare the relative value of expanding a wildlife refuge in the Florida Keys with the addition of acreage to an urban park near San Francisco, just as it is difficult to compare the value of supplementing federal holdings in the Sequoia National Park with the purchase of land next to the Antietam battlefield. But those are the types of decisions faced regularly by Congress in determining priorities for funding under the Land and Water Conservation Fund (LWCF).

Each year, Congress must decide how much money should be appropriated for the acquisition of public lands under the Land and Water Conservation Fund (LWCF). Four federal agencies are responsible for most of the federal government's land acquisition; those agencies hold approximately 626 million acres, nearly 28% of the United States. The Bureau of Land Management (BLM) accounts for 270 million acres, the U.S. Forest Service (USFS) manages 191 million acres, the U.S. Fish and Wildlife Service (USFWS) maintains about 89 million acres in 472 national wildlife refuges, and the National Park Service (NPS) manages 76 million acres in more than 360 units. (Other federal, state, and local holdings bring the publicly held land in the United States to slightly more than 40% of the total land area).

The four principal federal landholding agencies have differing mandates, histories, cultures, and criteria for choosing what lands to buy. Yet Congress must decide how to allocate the appropriations among those agencies, despite the difficulty of comparing different agency values. Since the LWCF was established by Congress in 1964, more than $3.6 billion has been expended by federal agencies to acquire lands, and another $3.2 billion has been made available in matching funds to the states for land acquisition, as well as for development of recreational facilities. In recent years, federal agencies have been active in land acquisition for conservation of recreation resources, historic preservation, establishment of wildlife and endangered species habitat, and other objectives. A recent shift toward greater recognition of ecological goals has in some cases strained federal-local partnerships and polarized groups seeking to preserve landscapes for competing ends.

To obtain assistance in evaluating the land-acquisition requests from the four principal landholding agencies, Congress asked the National Academy of Sciences to evaluate the land-acquisition criteria and procedures of BLM, USFWS, NPS, and USFS and to compare their methods with those of private groups, such as The Nature Conservancy (TNC), that are active in buying land for conservation. In response, the National Research Council appointed the Committee on Scientific and Technical Criteria for Federal Acquisition of Lands for Conservation.

The committee was asked specifically

- To review criteria and procedures by which NPS, BLM, USFS, and USFWS acquire lands for conservation;
- To assess the historic, public policy, and scientific bases of land acquisition criteria and compare them with nongovernmental organizations;
- To assess the effectiveness of the four federal agencies in preserving natural resources while achieving mandated public policy objectives of the agencies;
- To evaluate the extent to which agencies use objective methods, scientific knowledge, and systematic criteria in making their recommendations for acquiring conservation lands.

## CONSIDERATIONS FOR CRITERIA

To carry out its charge, the committee needed a framework to evaluate the criteria of the four agencies. Individual agency criteria and procedures are quite diverse; they include statutory mandates; well-documented, formal rules of practice; and informal procedures. Congress also has its own criteria, even if they are not formalized.

The differences between agency lists and the acquisitions funded by Congress have led some to suggest that political considerations override any systematic, objective criteria. Such suggestions seem to be based on a sense that analysis is somehow different from and superior to politics. But politics and analysis overlap in many ways, and politics is the expression—however imperfect—of group and individual interests. Those interests include complicated mixtures of perceptions, ideologies, economic interests, and even altruism. The diverse interests can result in conflicts, but they can also result in increased scientific information becoming available, as has occurred when scientists study areas to inform proponents or opponents of federal acquisition.

Thus, the committee needed to examine the various factors that make up land-acquisition policies and form the basis for the agencies' criteria. The committee concluded that workable criteria should

- Be consistent with agency missions;
- Contribute to achieving sustainability of renewable-resource bases, including biological and cultural diversity;
- Provide the basis for long-term planning, as well as the flexibility to take advantage of unexpected opportunities for land protection;
- Be feasible to administer and apply in a consistent manner;
- Respond to changing scientific knowledge, social values, policy, and public input;
- Clarify the significance of the proposed acquisition based on objective scientific and technical information;
- Distinguish among competing conservation values;
- Identify the distribution of social costs and benefits that would result;
- Be continually re-evaluated for performance and experience gained in their use.

In evaluating the criteria, the committee considered matters of process as well as the values and scientific issues that underlie the agencies' land-acquisition criteria.

## The Meaning of Conservation and Public Versus Private Land

The term "conservation" has many meanings; but most definitions share a concern for the future. Thus, the committee's working definition of conservation is the management of land resources to sustain their productivity in the long term and to avoid losses of valuable components. Conservation includes physical, biological, social, historical, agricultural, cultural, recreational, and aesthetic components. This report focuses on ecological and biological aspects of conservation, although the committee recognizes the importance of other public-policy objectives, such as cultural and historical preservation or equitable distribution of recreational opportunities. Further, the interests of landowners whose property is within public land boundaries ("inholders"), as well as the needs of nonresident populations must be considered. Priorities among objectives usually are value judgments, but all of the objectives are recognized by law and are integral components of conservation.

Private and public values often cannot be conveniently separated. As the interests of private and public landowners become more intertwined, the responsibilities of private managers change, and public managers become more alert to private property considerations. Conservation, including human needs, cannot be achieved through land acquisition alone.

## Scientific Bases for Conservation

Biological criteria for land acquisition have evolved with increasing understanding, from saving isolated areas for scientific observation to protecting biological diversity and the functioning of ecosystems and landscapes. But increased scientific understanding has also revealed that natural systems change incessantly. Studies of historical and prehistoric records make clear that the physical environment—including the climate—has been changing on a variety of time and space scales ever since life has existed on Earth. Organisms and ecosystems have changed as well. With this increased understanding has come general recognition that equilibrium models of the natural world often are not appropriate; dynamic models are required instead. This has profound consequences for conservation efforts. Conservation goals cannot be achieved in perpetuity simply and exclusively by strategic property set-asides. The model of an ecological system maintained in an undisturbed state is giving way to a dynamic view that seeks to protect the processes of natural dynamics and pays attention to size, shape, spatial arrangements, and connections in conservation lands.

Because many ecological processes occur over large time and space scales, conservation usually is more effective if conservation areas are connected. Populations of organisms in isolated patches are more likely to suffer local extinction than those in patches connected by corridors to other patches. Thus, integrated and complementary approaches are needed to identify critical unprotected areas, to connect fragmented and isolated habitats under the jurisdiction of separate agencies, and to meet a diversity of conservation needs, such as public access or wildlife migration. Making connections between lands, across agencies, and through planning is an important generic strategy.

## Tools and Techniques

The topic of this report—federal land acquisition—has many aspects. The term "acquisition" includes not only buying full title to parcels of land (acquisitions in fee), but also purchase of fractional interests (such as conservation easements), exchanges, and the exercise of eminent domain. Examination of NPS, BLM, USFS, and USFWS criteria suggests the emergence of several new property perspectives that could influence acquisition practice as it unfolds in the twenty-first century. One example is the use of methods that encourage transactions between willing sellers and buyers. Another is recognition that acquisition by public entities need not be limited to full-fee or top-dollar purchases but can include purchase of partial interests and other less-costly management mechanisms. Yet another example is the gradual blurring of distinct roles of private and public owners, where the responsibilities of private managers include public purposes and the responsibilities of public managers include private considerations.

A new conservation paradigm may come to rely more on property partnerships and management agreements between federal and local agencies and between public and private nonprofit conservation interests and private landowners than on strict federal ownership and control of the landscape. A notable example of such partnership is the "working landscape" model used in Europe, which focuses on the simultaneous protection of cultural and biological diversity and recognizes their coevolution and interdependence. These dual objectives are best obtained through mixed property systems that demonstrate versatility and practicality.

Finally, there is increasing recognition of the need for and power of social analysis in evaluating potential land acquisitions. There is never enough money available to meet conservation needs, regardless of whether full-fee acquisition, purchase of partial interests, or other management options are used. Social impact assessment can be a powerful tool in allocating limited funds among various acquisition options.

## CURRENT CRITERIA

National Park Service acquisitions can occur only within the boundaries of units authorized by Congress. Key considerations used to establish NPS acquisition priorities within individual park units are the primary purpose of the unit, land price escalation, legislative history, imminent threats, and protection of the unit. For any privately owned parcel, a land protection plan must identify the least federal interest needed to be acquired to achieve federal goals for that parcel. During the 1980s, NPS developed ranking criteria according to regional acquisition priorities; however, those criteria never were implemented.

The U.S. Forest Service, which manages the national forests for multiple use, uses a point system similar to that used by the Office of Management and Budget (OMB) (discussed below); projects must meet OMB's four minimum criteria. Other information gathered includes the type of area, priority within the region, acreage, location, price per acre, and total cost. Points are assigned based on whether the project meets specific needs in the forest plan, and on the OMB criteria.

The national wildlife refuge system of the U.S. Fish and Wildlife

## EXECUTIVE SUMMARY 7

Service comprises diverse lands, many of which were established and are managed for different purposes. In 1983, USFWS began to develop the Land Acquisition Priority System (LAPS), which defines five target areas: endangered species, migratory birds, significant biological diversity, nationally significant wetlands, and fishery resources. Separate criteria developed for each target area are derived from plans prepared under different authorities. LAPS has additional criteria common to all projects, including whether a proposed acquisition contributes to more than one target area and addresses threats to habitats. Regional offices develop lists of priorities and project proposals; those lists are compiled in a national data base to rank projects. Two final project lists are developed: one for migratory birds and one for endangered species.

The Bureau of Land Management first received general land-acquisition authority under the Federal Land Policy Management Act of 1976, which authorized the secretary of the interior to acquire lands or interests in lands by purchase, exchange, donation, or eminent domain. All BLM lands are to be managed on the basis of multiple use and sustained yield. Resource management plans, prepared for each management unit, contain detailed descriptions of lands available (primarily through exchanges) and lands that are desirable for acquisition. BLM's key objectives for wildlife are to acquire critical wildlife habitat and consolidate scattered tracts of land for efficient management of resources. Key objectives for recreation are to provide a diversity of recreational opportunities, provide resource-dependent recreational opportunities, manage and monitor essential resources, use land ownership and access to enhance recreational opportunities, and contribute to local economies by cooperating with tourism groups.

To rank acquisition priorities for overall LWCF appropriations, the Office of Management and Budget created a system that incorporates some aspects of NPS, BLM, USFS, and USFWS criteria, while emphasizing the administration's priorities, such as recreational opportunities near urban areas and wetlands protection. In identifying acquisition priorities for the federal budget, OMB ranks each agency's priorities according to a point system. Minimum standards to qualify for acquisition are

- Whether the proposed acquisition is within the boundaries of or is contiguous with an authorized unit;
- Absence of known health or safety hazards;

- Absence of opposition from current owners; and
- A limit of 10% of the purchase price for infrastructure expenses (e.g., costs of campsites and trails).

Points are awarded for different categories. The general categories are recreation and access (140 points), habitat and wetlands protection (120 points), cost minimization (70 points), threat of development (50 points), and protection of significant cultural and natural features (40). The assistant secretary for each agency can then add points for an agency's highest priority items.

## CONCLUSIONS AND RECOMMENDATIONS

The most successful acquisition criteria focus on specific purposes and well-understood policy goals, such as the USFWS criteria for migratory bird habitat or TNC criteria, which are based primarily on the goal of protecting biological diversity. By contrast, the OMB Land Acquisition Priority Procedure (LAPP) criteria lump the funding requests of several agencies into a single priority list submitted to Congress. In the formation of this list, agency missions often are compromised by evaluating noncomparable requests. Such a combined ranking system cannot address complex value comparisons.

Traditional acquisition practice has focused upon the evaluation of individual land parcels, apart from considerations of broader biogeographical and landscape patterns. Different conservation objectives among federal land-management agencies have led to a fragmented pattern of reserves selected mainly to protect specific resources. Specific conservation objectives do not necessarily address larger goals, such as the protection of entire ecosystems. Comparative evaluations of properties can be distorted grossly if they ignore the regional contexts and ecological dynamics of a particular land parcel or system of parcels.

Evaluations of land parcels on their individual merits and their transboundary planning potential are aided greatly by new technologies such as gap analysis and geographic information systems (GIS). Gap analysis and GIS are used widely in resource planning decisions, and their use should be encouraged and refined. Their usefulness, however, depends upon the adequacy of existing data and upon maps of ownership, inventories, population trends, and species distribution. Unfortunately, no comprehensive federal inventory of current landholdings in protected status is available to provide a basis for an interagency and regional perspective of land acquisition needs; therefore, it often is difficult to determine the success of criteria in meeting particular objectives. Better information is indispensable for improved acquisition decisions.

The involvement of nonprofit organizations has been a significant development in federal land-acquisition procedure in recent years. Those groups work as facilitators, innovators, dealmakers, and intermediaries between sellers and government buyers. Nonprofit groups often have rapid discovery and response capabilities that governments sometimes lack and can put together complex multiparcel transactions that cross agency boundaries and facilitate cooperation between public and private property owners.

Finally, although systematic planning and preparation are of great importance, they can result in overlooking some conservation opportunities; for this reason, a role for emergency acquisitions is justified. The Alaska purchase—called "Seward's Folly" at the time—illustrates how important seizing an opportunity can be in land acquisition.

After reviewing the current criteria used by NPS, BLM, USFS, and USFWS, the committee developed the following recommendations.

## Goals

*OMB and the four agencies should separate the current national ranking system into at least three categories: outdoor recreation resources, natural resources protection, and cultural resources protection.* Other categories may be needed, especially where Congress has designated portions of the federal lands for specific purposes, such as to protect specific kinds of resources, including wilderness areas, wild and scenic rivers, and historic and archeological sites.

*The criteria for conservation-land acquisition should be expanded to include landscape pattern analysis.* Such analysis generally includes land use and land cover data and measures factors such as patch characteristics, vegetation types, ecological trends, and hydrologic interactions with these resources. Land uses in an entire watershed should be considered in the design of reserves.

*BLM, NPS, USFS, and USFWS should prepare an overall strategic plan that identifies land-acquisition needs for establishing and protecting representative natural areas on federal lands that can provide scientific baselines for judging the effects of human actions on the natural environment.* Those needs should be recognized in the federal land-acquisition priorities.

*Agencies should use the widest possible range of land-protection strategies in formulating acquisition proposals.* That range should cover public ownership, land-use regulation, alternatives to fee-simple acquisition, exchanges, public-private and interagency arrangements, partnerships, cross-boundary planning, land banks, and other techniques.

### Procedures

*Agencies should develop and use long-term land-acquisition plans that can be used to identify priorities and opportunities for interagency cooperation.* Those plans should take into account regional conservation needs, as well as social effects of acquisitions on local landowners and communities, and they should provide a mechanism for public participation. The multiyear perspective of such plans would enable Congress to judge how well the agencies fulfill their missions and facilitate the evaluation of the cumulative effects of land acquisition.

*The federal land-acquisition program for conservation should have a common, interagency information base as part of a systematic approach to achieving its goals (e.g., protection of biological diversity, wild and scenic rivers, cultural heritage, public recreation).* This information base should enable the land-management agencies and Congress to determine the extent to which conservation needs are being met and to identify gaps in meeting these needs. Such information should be expanded and assembled for all four agencies in a common GIS. The agencies should continue to refine and expand their applications of gap analysis and GIS. Data gathering should be improved, extended, and directed with a view towards applications in gap analysis and GIS. When feasible, social impact analysis should be used.

*Federal acquisition criteria should distinguish national from state and local criteria for outdoor recreation and other conservation needs.* NPS, BLM, USFS, and USFWS should consider public outdoor recreation opportunities and conservation needs and resources of state, local government, and Indian tribal lands in federal land-use plans.

*Private land managers should be encouraged to achieve goals that previously have been achieved primarily through acquisition by establishing partnerships and other means of cooperation.* This emphasis is needed because conservation needs cannot be satisfied through public land-acquisition programs alone.

*For long-term planning and consistent adherence to a set of criteria, the LWCF needs adequate and predictable funding.* National planning should be attentive to local planning. National criteria should be tied to criteria used in local land-use plans and should give weight to congressionally designated areas.

*Congress should develop effective mechanisms, such as providing discretionary LWCF funding, for dealing with emergencies and unexpected opportunities.* This would permit the secretaries of the Department of the Interior and the Department of Agriculture to take advantage of unexpected opportunities or respond to unwelcome threats to resources.

*Acquisitions should periodically be assessed retrospectively to determine if the purposes for acquisition have been achieved.* Criteria also should be periodically re-evaluated in the light of changes in holdings, biological resources and climate conditions, demographics and public recreational needs, scientific knowledge and data bases, concepts of conservation, as well as changing concepts of property in the service of conservation goals.

*Agencies should continue to take advantage of the ability of the nonprofit organizations to act swiftly to secure properties until an agency can acquire them.* This can be done while ensuring that federal acquisition priorities guide the process and that the specifics of the transactions be in accordance with federal guidelines that control dealing with nonprofit organizations. Nonprofit organizations historically have played a vital role in enabling government agencies to carry out acquisition programs expeditiously and effectively.

# Setting Priorities for
Land Conservation

# 1

## Introduction

Established in 1964, the Land and Water Conservation Fund (LWCF) is the primary source of federal funds for acquiring land for conservation. Since 1964, more than $3.6 billion has been spent by federal agencies to acquire land; another $3.2 billion has been available as matching funds for the states to acquire land and to develop recreational facilities. The purposes for which acquisitions can be made have been expanded to include other objectives, such as establishment of wildlife and endangered species habitat. That shift sometimes has strained federal-local partnerships and polarized groups seeking to preserve land for competing purposes.

Each year, Congress must decide how much should be appropriated for land acquisition and how the amount should be allocated among the various federal agencies and the states. Concern about how LWCF funds are distributed by federal agencies and how the different agencies choose acquisitions prompted Congress to ask the National Academy of Sciences to evaluate the land-acquisition criteria and procedures of the four agencies that are responsible for the bulk of land acquisition—the Bureau of Land Management (BLM), the Fish and Wildlife Service (USFWS), the National Park Service (NPS), and the Forest Service (USFS)—and to compare their methods with those of private groups, such as the Nature Conservancy. In response, the National Research Council appointed the Committee on Scientific and Technical Criteria for Federal Acquisition of Lands for Conservation in the Board on Environmental Studies and Toxicology. The committee was asked specifically

- To review criteria and procedures by which BLM, USFWS, NPS, and USFS acquire lands for conservation;
- To assess the historic, public policy, and scientific bases of land-acquisition criteria and compare them with nongovernmental organizations;
- To assess the effectiveness of these federal agencies in preserving natural resources while achieving mandated public policy objectives of the agencies;
- To evaluate the extent to which agencies use objective methods, scientific knowledge, and systematic criteria in making their recommendations for acquisition for conservation.

The committee was composed of members with expertise in public land and wildlife law and policy, national resource economics and management, land use, sociology, conservation biology, ecology, hydrology, and watershed management as well as experience with land-acquisition practices and transactions.

Over the course of its deliberations, the committee heard presentations from national and regional representatives of the four land-management agencies, as well as representatives from some nonprofit conservation organizations and organizations that represent interests affected by land acquisition, such as the Farm Bureau. The committee also had opportunity to garner perspectives from local communities and governments. (Appendix A is a list of persons who made presentations to the committee and participated in discussions of its task.)

The committee evaluated the land-acquisition criteria of BLM, USFWS, NPS, and USFS in the context of agency missions, the dynamic nature of ecosystems and landscapes, human conservation needs, and changing social objectives. The committee considered patterns of land-ownership, the many meanings of conservation, and the relationships among government agencies, nonprofit organizations, private landowners, diverse populations, and other interested parties that influence the acquisition process.

This report emphasizes primarily the ecological and biological aspects of conservation while recognizing the importance of social objectives, such as cultural and historical preservation and equitable distribution of recreational opportunities. Priorities among objectives are value judgments beyond the scope of the committee's charge, but all of the objectives above are recognized by law and are integral components of conservation.

## LAND-ACQUISITION AGENCIES

The four federal agencies reviewed have separate, somewhat complementary, missions. All are responsible for managing large areas of land, most of which is rural; but overall, the lands present a wide range of conditions and opportunities for meeting biological and human needs. Land acquisition is one tool used by Congress and the agencies to fulfill their missions. The four agencies are responsible for the following:

- The National Park Service is responsible for managing the national parks and monuments, as well as cultural and historic sites, constituting more than 350 units on 76 million acres, the largest part of which is in Alaska (54 million acres) and the West. These lands have natural, historic, and scenic features and provide public recreation opportunities.
- The Fish and Wildlife Service maintains the national wildlife refuge system, which has some 89 million acres (approximately 60 million in Alaska) in 472 units; protects nesting and migration habitat for migratory birds; provides habitat for endangered and threatened species and for other wildlife; offers recreational opportunities for the public; and permits other uses of the refuges where compatible with the primary wildlife purposes.
- The Forest Service manages the national forests and grasslands, which together account for about 191 million acres of forest and rangelands throughout the United States. USFS lands are available for a wide range of commodity uses, recreation, and resource protection.
- The Bureau of Land Management manages the federal resource lands, about 270 million acres, almost entirely in the 11 contiguous western states and Alaska. Much of those lands is arid and is located in small, scattered tracts. These lands are used for recreation, commodity production, and resource protection.

Many of the individual units of these federal land systems are solid parcels of land, but many also have significant gaps in the land base. And units are created from time to time to meet newly recognized needs. Land acquisition is used to obtain privately owned land within existing units as well as to expand the boundaries of federal land systems and to connect existing systems.

Some federal lands are controlled by agencies other than the four considered in this report. After the departments of the Interior and Agriculture, the Department of Defense controls the third largest amount of federal land—3.9% (26 million acres) (GSA, 1989). Federal lands controlled by agencies other the BLM, NPS, USFS, and USFWS might have an important part in the management of wildlife and recovery of threatened and endangered species. Several colonies of the endangered red-cockaded woodpecker, for example, are located in federal forests on military bases in the South.

## GUIDELINES FOR CRITERIA

To evaluate criteria for land acquisition, some standards are needed by which the value and utility of criteria can be measured (Fink, 1991).

Acquisition criteria may appear in an agency's organic law, as in BLM's Federal Land and Policy Management Act; in general acquisition statutes, such as the Condemnation Act and the Uniform Relocation Assistance and Acquisition Policies Act of 1970; in a wide range of programmatic laws, such as the Endangered Species Act, the Wild and Scenic Rivers Act, and the National Trails System Act; and in a host of project-specific or unit-by-unit laws—the National Park Service alone is subject to more than 40 laws of this sort spelling out the practices of eminent domain.

Land-acquisition criteria also emerge as a result of explicit deliberation within and among the agencies (e.g., the USFWS Land Acquisition Priority System (LAPS) Application Manual, the USFS Land Acquisition Handbook, and the Office of Management and Budget (OMB) Procedure for Compiling Federal Land Acquisition Priority List); in general planning documents, such as BLM's *Wildlife 2000* and *Recreation 2000* documents; as well as in regional planning endeavors, in action-specific impact statements, and in species recovery plans prepared in accordance with the Endangered Species Act of 1973.

Land-acquisition criteria may be detected in the common practices of agencies. NPS, for example, does not often attempt to acquire partial interests (e.g., easements or development rights). BLM and USFS land exchanges are constrained by the complexities of land-exchange transactions. Congress itself has certain procedures to bring political considerations into the process and has ultimate control of land-acquisition priorities through the appropriation process. Any number of unspoken rules and hard-to-identify practices govern contacts between government agencies that acquire land and the people who own it.

Comparing such diverse criteria is extremely difficult, because each participant has different goals and priorities, different sets of statutory and administrative constraints, a different notion of what constitutes conservation and lands appropriate for conservation, and different modes of acquiring them. The criteria themselves are variable, addressing different small pieces of a large picture.

The most successful acquisition criteria examined by the committee were constrained and focused by well-understood policy goals. It is one thing to look for the best acquisition for a specific purpose, e.g., to maintain an endangered species, protect cultural artifacts, or provide hiking or bicycling opportunities. It is quite another to try to achieve those goals simultaneously through composite criteria intended to determine the highest priorities among the goals.

A private group, such as The Nature Conservancy, can pursue its goal of protecting "natural" places without the homogenization of purpose that can hobble the acquisition programs of public agencies. Thus, private groups can be innovative in defining goals and methodologies (Rousch, 1991) without attempting to satisfy multiple purposes and constituencies. Among federal land-acquisition agencies, the advantages of singularity of purpose and coherence of aim are found in USFWS Land Acquisition Priority System (LAPS) criteria (see Chapter 3), which are an excellent vehicle for identifying migratory bird habitat within the United States.

## Basic Considerations

The committee identified four important considerations that need to be evaluated in determining appropriateness of criteria.

*Conservation of sustainability:* Criteria should contribute to sustainability of renewable resource bases within a particular region. Conservation for sustainability includes the consideration of cultural and biological diversity and the processes that renew or perpetuate them. The goal of sustainability is beyond the capacity of any single program, discipline, or organization to achieve, but land acquisition can play an important role that complements other programs.

*Fulfillment of agency purpose:* Criteria should further the missions of the agencies and help to achieve their land-acquisition goals. If the use of criteria leaves certain agency missions unaccomplished or largely neglected, the criteria are inadequate.

*Clarity:* Criteria should lend themselves to ease of administration and consistency of application. Criteria that are either so complex as to defy understanding or so subjective as to permit manipulation are unsatisfactory.

*Accommodation of variation and change:* Any land-acquisition priority scheme intended to function over time confronts a variety of policy challenges, including continuity despite changing conservation needs and values, the capacity to plan for and implement comprehensively while responding to opportunities, and the need to meet a variety of specific and often competing goals.

Scientific knowledge, values, and other circumstances change over time; useful criteria must be flexible enough to respond to these changes. For example, adequate protection of biological diversity has led to various proposals to restore or protect endangered habitats by assorted policies, including land acquisition. NPS, USFS, USFWS, and BLM all pursue conservation strategies that rely upon continuous collection of evidence and ongoing evaluations that give presumptive weight to historical practices.

Written criteria and land-acquisition policies change in the course of agency practice and administrations. The missions of the agencies themselves have been reshaped and modified over time, and land-acquisition practice often is the subject of rapid-fire legislative instruction, program by program, and project by project.

The Land and Water Conservation Fund Act has been modified in response to legislation to protect endangered species, establish wildlife refuges, and protect wetlands. LWCF beneficiaries have changed also—BLM was not an original beneficiary of the LWCF, but it now is a major recipient. State and local governments have not fared well throughout the many changes to the LWCF, although their needs were considered by the original drafters of the LWCF as being at least as urgent as those of the federal agencies (Glicksman and Coggins, 1984).

In addition, the lands acquired and public expectations regarding them change over time. For example, the USFS acquired forested land in the eastern United States for timber management. The land had been cut over, and during the decades when no harvesting was possible on the lands, the agency's custodial management was noncontroversial. People's expectations about the land changed to favor hunting, fishing, and recreation. When the timber was once again of merchantable size, the timber sales program was not acceptable in some cases to the local population. Land that had been acquired primarily for timber management had become important for other conservation purposes.

The physical and biological environment changes constantly. Suitable habitat for a species requiring slightly over-aged timber stands will not meet that criterion forever. And climate change probably will bring physical and biological changes on large scales and short time frames, as it did in earlier times.

Changes such as those described above interact with strategies of acquisition in subtle ways. The most reliable way to achieve permanent dedication of a parcel of land to a stated purpose is to acquire the entire parcel and all related interests (a "full-fee" acquisition). But changes in the behavior of a protected species, for example, or in climate or the scientific understanding of habitat needs, might warrant protection with a broader reach.

## Planning Versus Opportunity

Formal acquisition criteria permit planned decision-making; nonetheless, unanticipated opportunities arise and disappear quickly, and criteria must be responsive. An acquisition program should adhere to a standard planning model that includes identification of unambiguous goals, specification of alternatives, and selection of an option that best advances identified goals.

Planning on federal lands is expressed in a variety of ways, including land-use management plans, land-protection plans, and environmental impact statements. Such formal planning and deliberate choice is reflected also in the acquisition criteria. Land-acquisition decisions involve a comprehensive search for prospects, careful evaluation of the costs and benefits associated with serious nominees, weighing of alternative choices, and selection of the best candidates measured against criteria. But opportunism still has a role in the land-acquisition policies of a nation that was the reluctant recipient of the Louisiana Purchase and called the purchase of Alaska "Seward's Folly."

Acquisitions involving opportunistic land exchanges can be almost entrepreneurial endeavors, partly because land acquisition through exchanges presents issues and functions that generally do not conform to established systems for setting priorities. An illustration of the opportunistic nature of land acquisition is presented by the recent collapse of savings and loan associations. In short order, the federal bailout law brought into the hands of the Federal Deposit Insurance Corporation and the Resolution Trust Corporation many valuable properties that included sensitive wetlands and estuarine areas (Frederick, 1991). Abandoned railroad rights of way also have provided recent fleeting opportunities for other uses. Acquisition practice must be responsive to sudden and unexpected opportunities of this sort.

Nonprofit organizations often have exploited sudden opportunities to the benefit of the nation. Groups and entities with diverse concerns can fulfill functions that governments cannot—overcoming financial limitations and red tape, dealing with owners who do not wish to bargain with governments, responding to sudden crises and fleeting opportunities, satisfying the private owners' desires to sell quickly, arranging complicated transactions, and providing services for the agencies that eventually will manage the land (Montana Land Reliance/Land Trust Exchanges, 1982).

## Acquisition and Alternatives

In public land law, land acquisition includes full-fee purchase and condemnation by eminent domain; acquisition of lesser interests, such as easements, rights of way, and life estates; and other means, such as sales, exchanges, gifts, and bequests. Beyond this, mechanisms used to manage the federal lands include acquisition, disposition, and numerous use and allocation decisions. It is important to underscore the essential continuity and interconnection among those mechanisms. Response to allocation of resources leads to changing public demands, which can lead to new acquisition, management, and disposal decisions. Acquisition is one means to accomplish management goals, and it can be used as an alternative to other forms of management or as a supplement to regulation. For example, advocates of vigorous enforcement of Section 404 of the Clean Water Act to restrict development in wetlands believe that acquisition should be used to supplement regulation (Houck, 1988).

But acquisition is not a substitute for regulatory land-use restrictions. Circumstances could arise in which an aggressive acquisition strategy aimed at wetlands or endangered species habitat, for example, could work against regulation—by driving up expectations and demands of landowners that their land be acquired by the government, an expectation that might never be satisfied in fact. Strategic acquisitions, however, have an important role to play across the spectrum of management activity.

Broader experimentation with less-than-fee acquisitions is warranted for two reasons in particular: future management aims likely will be ecosystemwide, and they will reach across public and private lands. The Nature Conservancy underscores the implications of this approach:

> As The Nature Conservancy turns to preserving whole landscapes, it will necessarily pay attention to human needs. For example, neither the Conservancy nor anyone else will be able to buy all the critical land needed to protect the 100 miles of the Sacramento River habitat targeted by the Conservancy's California office. It will take voluntary agreements with landowners, cooperation with other nonprofit groups and local land-use regulations. It will take sophisticated coordination with an array of state and federal agencies (Rousch, 1991).

Indeed, the ability to forge amicable relationships with resident populations might be the dominant factor for success or defeat of parklands and preserve initiatives nationally and internationally (West and Brechin, 1991).

The distinction between the rights of public and private landholders is becoming less obvious. Public and private ownerships are evolving and converging, with private rights emerging in the public lands, and public obligations appearing in the private sector. Private property normally is subject to regulatory land-use restrictions to protect public values, while acquisition policy is shifting from what might be called a perspective of ouster to one of accommodation.

Much of the legal structure of federal land acquisition is based on condemnation, classically exemplified by the interstate highway system and urban renewal programs, which were completed by taking the properties of millions of Americans. Policy based on condemnation has many implications, including a low tolerance for inholders and unwilling sellers; indeed, the popular perception of condemnation proceedings presumes an unwilling seller versus the government. But condemnation can be undertaken between a willing seller and the federal government In a prominent example of accommodation, the committee was advised that NPS has moved gradually from a policy of attempting to remove all inholders to a policy of removing only those inholders whose uses are incompatible with management objectives. And a keystone of OMB acquisition policy is the presence of a willing seller (see Appendix B).

Several factors contribute to the shift in strategy. Experience has shown that the purposes of LWCF acquisitions can be achieved largely within the framework of a willing seller. Properties acquired for recreation, wildlife habitat, cultural, or scenic protection do not always depend upon eradication of existing uses and removal of occupants. In some instances, occupant values even reinforce such objectives and can be an asset rather than a liability.

## THE INFORMATION GAP

Determining the ownership of conservation land is difficult. The committee was unable to find any comprehensive source of information regarding what privately held lands the federal government considers desirable to acquire for conservation and what lands it already holds for conservation. More pertinent, perhaps, is that the committee could not identify what lands the federal government has acquired for conservation. Even with the narrowest possible definition—lands acquired with LWCF monies—no comprehensive source is available to learn where lands have been acquired, by which agency, at what cost, and for what purpose, although such information is available in theory. The multitude of partial holdings (e.g., conservation easements that restrict land uses), which are difficult to define and map, also obscures basic information about federal land holdings.

## VALUATION CHALLENGES

Any criteria for prescribing priorities for public policy actions are fraught with difficulties in comparing values, even within constrained contexts, such as those of USFWS LAPS criteria. Comparing values across the spectrum of resource use and geographical happenstance is even more difficult.

One common technique to compare incommensurables is quantification. But even the most robust quantification schemes cannot eliminate the need for decisions based upon experience, reasoned judgment, intuition, and common sense. Importance of diverse values cannot be expressed in an ordinate arrangement.

Land-acquisition criteria should deal openly and explicitly with differences in value between the parcels competing in the priority scheme and convey adequate and meaningful information on value to the interested public and to Congress. The result of applying the criteria should be an efficient allocation of limited funds, providing the most gain to the nation for its investment. The criteria also should achieve widely shared goals that are anchored firmly in law.

Ultimately, land-acquisition decisions affect many people and organizations, including property owners and local communities. The criteria used to guide land-acquisition priorities should reflect this and seek to treat those affected interests fairly. The criteria should invite a continuing re-examination of their premises in light of their performance and the experience gained in their use.

## REPORT ORGANIZATION

The committee focused on technical and scientific criteria for land acquisition and on the way potential acquisitions are ranked at the national level. The committee did not view its role as one of evaluating goals for land acquisition that have been set by Congress or as one of assessing the validity of the agencies' missions.

The committee reviewed the changing policy framework within which federal land acquisition takes place. Thus, it considered the effects of shifting land uses and population change on maintaining biological diversity as well as the social and cultural dimensions encompassed within the modern understanding of conservation.

This report starts by placing federal land acquisition in a historical and thematic context (Chapter 2). It describes the extent of the federal lands, their relation to other lands, and the broad purposes of federal land acquisition.

Chapter 3 examines the organizations and interests that directly affect and are affected by federal land acquisitions. The missions, land-acquisition programs, and the operations of NPS, BLM, USFS, and USFWS are described. The allocation of land-acquisition funds by Congress is discussed, as are the roles of various interested parties, including those who hold private land in or adjacent to the federal lands. The roles of private groups that also protect conservation lands and assist the federal agencies in fulfilling their missions are described.

Chapter 4 examines the social dimensions of land acquisition, including cultural issues that affect land policy and conservation and human needs. Chapter 5 addresses biological aspects of conservation and technical procedures for establishing priorities. Chapter 6 describes the criteria of nonprofit organizations for land assembly.

Chapter 7 describes the various techniques and tools used in acquiring land and interests in land, as well as examples that demonstrate the complexity of acquisition strategies and transactions. Factors that determine which techniques are most applicable are described. Chapter 8 is the committee's evaluation of the Office of Management and Budget's Land Acquisition Priority Procedure.

The final chapter of the report presents the committee's evaluation and recommendations. The current process for setting acquisition priorities is evaluated and recommendations are made for increasing the effectiveness of the process for establishing land-acquisition priorities.

# 2

# Public Land, Private Land: An Overview of Ownership and Its Management Challenges

In an era when privatization and deregulation have widespread appeal, proposals to extend public ownership of land must prove that such ownership is in the public interest. Land acquisition decisions must account for conservation incentives and disincentives confronted by private and public land managers, as well as other factors (e.g., population growth, changing occupational structure, and new institutional conservation mechanisms).

## CONSERVATION: A TERM OF MANY MEANINGS

Although the history of conservation has been characterized by conflict over the meaning of the term, one element has been common to nearly all sides. Ciriacy-Wantrup (1985) noted that the theme underlying various uses of conservation is a concern for the future. He argued that conservation is concerned with the "intertemporal distribution of resource use." Thus, land conservation could be defined as the management of land resources to sustain their productivity in the long term and to avoid losses of valuable components. Dana and Fairfax (1980) warned that conservation can be defined only in the context of particular times and places. Its meaning might even vary with the purposes of the speaker. A brief examination of the history of the concept can help to circumscribe the range of meanings as well as to identify prominent and common elements of conservation thinking.

## The Nineteenth Century: Preservationists and Progressives

During the late nineteenth century, conservation was a contested term used by those advocating preservation of vast tracts of land for natural and intrinsic values, as well as by those who supported orderly development of natural resources (Hays, 1959; Fox, 1986). Preservationism is apparent in the writings of Ralph Waldo Emerson, Henry David Thoreau, and John Muir, who saw wilderness as a place of escape, re-creation, and closeness to nature. Many preservationists believed that as the country became increasingly urban, experience with nature was necessary to rebuild the American character (Gilligan, 1953). But preservationists also saw wilderness being lost to private exploitation. They argued that government should act to reserve portions of the public domain to ensure that present and future generations would have the opportunity to observe virgin forests, mountain meadows, and other features of nature. Some, such as John Muir, argued that nature had a right to exist that was independent of society's needs (Fox, 1986).

Scientists also played a crucial role in the emergence of the preservation movement (Fleming, 1972). The linkage between aesthetic appreciation of nature and observation and explanation of natural processes provided a foundation for arguments to preserve some lands for scientific exploration, public visitation, and nature preservation.

Another preservationist argument was what historian Alfred Runte (1990) has called monumentalism—that preservation of visually spectacular landscapes would reflect the riches of America. The United States of the nineteenth century lacked the cultural achievements of Europe, but it had tremendous natural wonders. Preservation of these landscapes provided a favorable basis for comparison with Europe as well as an additional base for economic development.

Preservationists used a variety of arguments to promote the reservation of vast tracts of land from disposition under the public land laws. Preservationists were influential in the establishment of forest reserves, national monuments, and national parks through the early years of the twentieth century.

Those who advocated development of natural resources also exercised considerable influence on federal land policy. The progressive conservation movement, led by Theodore Roosevelt, Bernhard Fernow, Gifford Pinchot, and much of the federal scientific establishment, attempted to bring about efficient use of natural resources through large-scale management systems directed by professional resource specialists (Hays, 1959). Progressive conservationists argued that professional managers would be free from the corrupting forces of politics; thus, resource management would be based on scientific principles and the public welfare. The broader goal of these conservationists was "foresight and restraint in the exploitation of the physical resources of wealth as necessary for the perpetuity of civilization and the welfare of present and future generations" (Wells, 1970). This concept prevailed in federal conservation agencies through the first half of the twentieth century (Hays, 1959, 1987; Fleming, 1972; Worster, 1985).

## The Early Twentieth Century: Expansion of Resource Management and the Movement for Wilderness Preservation

Conservation strategies and techniques were developed and extended to a range of natural resources during the 1920s and 1930s (Hays, 1987). In the Forest Service (USFS), conservation meant management of forests to yield crops of timber in perpetuity. Foresters would promote conservation by practicing sustained yield on the national forests and by providing a knowledge base and technical and protection services to make scientific forest management possible on private lands (Greeley, 1972; Steen, 1991). Water conservation, as practiced by the Bureau of Reclamation, meant construction of large dams to provide irrigation, flood control, and electric power (Reisner, 1986). Water projects facilitated human use of rivers so that water would not be "wasted" by flowing out to sea. Moreover, waterways were regulated to minimize the damage that flooding rivers could cause. During the New Deal, the Soil Conservation Service encouraged farmers to adopt practices that would minimize soil erosion. Conservationists also developed wildlife policies that emphasized limited hunting seasons, bag limits, predator control, and establishment of refuges (Worster, 1985; Fox, 1986; Dunlap, 1991).

Perhaps the most distinctive development in conservation policy during the 1920s and 1930s was the designation of wilderness areas in the national forests. Several USFS officials, including Aldo Leopold and

Robert Marshall, argued that the agency should preserve areas where natural resources would not be exploited (Gilligan, 1953). Leopold argued that wilderness was a resource that could be used to preserve the best parts of American culture. He believed that by bringing Americans into regular contact with the frontier that shaped their culture, qualities such as individualism mediated by cooperativeness, intellectual curiosity combined with pragmatism, and independence would survive (Leopold, 1925). Thus, for Leopold, an important criterion for the designation of wilderness was proximity to population centers.

Leopold also offered an argument for wilderness preservation based on scientific and resource management needs (Leopold, 1933). His observations of wildlife and work in game management coupled with his understanding of ecology led him to suggest the need for preserving areas to provide opportunities for scientific study of undisturbed nature and thus to improve understanding of the consequences of manipulating natural systems.

Like Leopold, Marshall (1930) emphasized the cultural benefits of wilderness. He argued that confrontation with the wilderness built physical and mental stamina, provided intellectual stimulation, and offered a distinctive beauty. Numerous other authors have extolled the values of wilderness, and nearly all have emphasized the enhancement of human character or spirituality associated with confronting nature.

Despite such efforts, the progressive conservationists continued to dominate federal policy. Resource-management schools trained cadres of professional foresters and wildlife and range managers to staff federal agencies. But during the 1950s, a small core of citizens' groups was increasingly successful in gaining public support for wilderness preservation and protection of nature for nonutilitarian ends. They depicted federal resource-management policies as destructive of nature and defined conservation as preservation of land untouched by human development. Their perspective gained support from a growing segment of the public, and during the 1960s, that support was reflected in congressional passage of wilderness, wild and scenic river, and endangered species legislation.

## Conservation in the Age of Environmentalism

The rise of the environmental movement during the 1960s gave impetus to even more widespread public acceptance of preservationist thinking and once again made conservation a contested term in public policy debates (Dana and Fairfax, 1980; Hays, 1987). Advocates of resource exploitation and of preservation both claimed the title of conservationist. This contest over federal land, water, and wildlife policies continue to be waged today. Competing interpretations of conservation were reflected in battles over clearcutting, wilderness designation, old-growth and ancient forest preservation, and endangered species protection.

One major commission provided a forum for conservation debates and shaped subsequent institutional responses: In 1958, Congress created the Outdoor Recreation Resources Review Commission (ORRRC). The 1963 ORRRC report recommended the creation of federal grants-in-aid to states to support recreation planning, land acquisition, and site development. Other proposals included adoption of wilderness legislation, a wild rivers program, and more efforts to support urban recreation (Dana and Fairfax, 1980).

The ORRRC report was well received by both major parties, and Congress acted on many of its proposals. In 1964, Congress passed wilderness legislation and created the Land and Water Conservation Fund (LWCF). A primary purpose of LWCF was to support federal, state, and local acquisition of lands for outdoor recreation resources. In addition, LWCF was used for state recreation planning and state and local construction of recreation developments (Dana and Fairfax, 1980). The LWCF was largely noncontroversial; however, the Wilderness Act, along with subsequent preservationist legislation, provided a stage for numerous later conservation battles.

In subsequent years, scientific, institutional, and political developments reshaped many of the dimensions of the conservation scene. Private land trusts or conservancies became increasingly important in conserving land (Bremer, 1984; Elfring, 1989; Morine, 1990). These organizations are private, nonprofit corporations that protect specific parcels of land through direct transactions, including acquisition by purchase or donation of land or interests in land. Purposes range from preserving wildlife habitat to establishing community gardens (Bremer, 1984). Although the objectives of these organizations reflect various understandings of conservation, the common theme among them is to save land from development—particularly urban development—that would be inconsistent with the goals of the organization.

Another trend was the movement for deregulation, privatization, and generally less government, sometimes called the "Wise Use Agenda" (Gottlieb, 1989). "New resource economists" were at the front of this movement, and they advocated alienation of natural resources from the public to the private sector (Anderson and Leal, 1991). They argued that private ownership and management would yield more efficient and equitable outcomes than would public ownership and management (Salazar and Lee, 1990). Conservation goals could be achieved by defining private property rights to natural resources and permitting their exchange in competitive markets (Anderson, 1983). Private owners would have incentives to maintain the productivity of resource lands while ensuring that they produced desired goods. The new resource economists, together with the "sagebrush rebels" (who advocated that public lands be ceded to the states), mounted an attack not just on federal acquisition but on continued federal ownership of conservation lands.

During the 1950s and 1960s, resource economists focused their efforts on modeling socially efficient rates of resource exploitation (Ciriacy-Wantrup, 1952; Gordon, 1954). They calculated rotation ages for timber stands, maximum sustainable yields for fisheries, and the net benefits of water developments and related them to vesting people with private-property rights in those commodities. During the 1970s and 1980s, many resource economists turned their attention to valuation of noncommodity resources, such as recreation and biological diversity (Randall, 1987; Montgomery and Brown, 1991). Resource economics, which had been the social science arm of progressive conservation, increasingly addressed issues raised by conservation biology.

Conservation biology contributed to changing understandings of conservation during the 1970s and 1980s. Conservation biology is a multidisciplinary field (Soulé, 1985; Noss, 1991) that draws on research in ecology, population biology, and biological geography to preserve biological diversity. Conservation biologists have developed management strategies to conserve diversity at the levels of gene pools, species, and ecosystems.

Many conservationists and environmentalists have used conservation biology as the scientific basis to advocate expansion of the wilderness system and to support biological diversity. Advocates include radical and mainstream elements of the conservation movement. They have argued that conservation should not be pursued solely for human ends; species and ecosystems should be preserved in their own right. Moreover, an important criterion in the creation of wilderness preserves should be to maintain biological diversity.

Progress in the natural sciences often precipitates changes in popular conceptions of conservation, but changes in public policy do not necessarily follow immediately. For example, although the Endangered Species Act offers a means to achieve biological diversity at the species level, there is little statutory basis for a conservation strategy aimed at protecting biological diversity at either the level of genetic material or ecosystems.

## Natural Sciences and Conservation Efforts

Developments in the physical and biological sciences influenced conservation thinking. This section examines whether the criteria used to acquire lands for conservation reflect current scientific knowledge about the requirements for the conservation of natural resources.

The key concept of sustainability attracts considerable attention and expanded application today. Each of the four agencies reviewed by the committee has mandates involving sustained production or maintenance of some element of the resource base. Language such as "in perpetuity," "to leave unimpaired," "long, wise use," and "multiple-use sustained yield" imply the maintenance of the land and resource base far into the future.

Early discussions of conservation pitted preservationists against progressive conservationists. The progressive conservationists developed the science of resource management and resource economics but often ignored the ecological factors affecting resource use and management, that is, a more comprehensive perspective on the environment, its ecosystems, and interactions. Lands for conservation in America were acquired in part to protect specific resources without regard to the broader ecological issues that affect sustainability. The result was a fragmented pattern of reserves for protecting specific resources.

The emergence of the field of wildlife biology and information about species distribution, threats, and extinction contributed to development of endangered species legislation. Focusing on the protection of single species without emphasis on ecosystems has led to land acquisition criteria that do not adequately address protection of species within an ecosystem or landscape context. For example, water resources have not been considered in land acquisition decisions for conservation until recently. And natural-resource management has suffered from some erroneous concepts fostered by a fragmented view of resources. Numerous other examples demonstrate instances in which lack of knowledge of the resource base contributed to conservation policies that were not comprehensive in approach.

Equally disturbing, however, is that essential works of science that might have guided land acquisitions, such as John Wesley Powell's work on arid lands (Powell, 1879), were ignored with respect to settlement, and policies were developed that circumvented resource-limiting factors. As the turn of the century approaches, it is useful to take stock of the sciences and their role in shaping conservation thinking. Moreover, important developments and perspectives in several related scientific fields in the past several years will contribute significantly to long-term conservation thinking and fulfillment of agency mandates in the next decade.

## Elements of Conservation and Its Institutionalization

Conservationists have tried to achieve an array of objectives using numerous managerial strategies and a small set of institutional arrangements. Managerial strategies have included sustained production of commodity resources using an agricultural model, active management of reserves either to maintain a particular type of landscape or to protect biological diversity, and holding wilderness areas that are not directly managed. Progressive conservation continues to influence federal land management agencies, and the agricultural model of conservation for commodity production is embodied in much of what those agencies do. However, conservation biology and related disciplines are gaining support; hence, active reserve management is receiving more emphasis. Federal ownership and management has been the predominant institutional arrangement for land conservation, but this situation is changing as state and local governments and private entities undertake more conservation activities.

In a broad sense, acquisition includes obtaining partial interests ("less-than-fee" interests) in land. Federal acquisition is not the only available conservation tool. Acquisition by state and private conservation organizations may become increasingly important, and regulatory and tax authorities can be used to achieve conservation ends. Efforts to analyze conservation policy will be most useful if they address the diverse set of purposes that conservation has served and might serve as well as the broad array of instruments for achieving conservation goals.

Conservation efforts largely have reflected interests in different uses of land. The issue that has divided conservationists is the purpose to be served. Thus, one way to analyze conservation policy options is to identify and characterize the resources or goods associated with the various meanings of conservation. Social scientists have developed conceptual models that link characteristics of resources to appropriate institutional arrangements for their management (Ostrom, 1990; Salazar and Lenard, 1992). Those kinds of models can be used to consider how particular conservation goals can be achieved.

## PUBLIC AND PRIVATE LANDOWNERSHIP

### Public Landownership in the United States

The use of criteria to establish priorities for land acquisition is limited by the lack of a comprehensive source of information on public and private landownership that provides an overview of lands that are protected or suited for conservation purposes. All of the land the United States has acquired on the North American continent previously was owned by other nations and Indian tribes. Federal acquisition came about through a combination of purchase and conquest (Coggins and Wilkinson, 1987) (Table 2-1 and Figure 2-1). Until passage of the Indian Appropriations Act in 1871 (U.S. Stat. 544, 566), treaties were negotiated to transfer Indian title to the public domain and reserve certain lands (held in trust by the federal government) for the tribes. Much of the reserved lands became public domain and was transferred to non-Indian ownership after passage of the Dawes Severalty Act of 1887 (also known as the General Allotment Act). That act assigned 160-acre allotments to individual Indians; the remaining reserved lands—more than 60

TABLE 2-1 Acquisition of U.S. Territory from Foreign Nations

| Land | Date acquired | Total area (land and water) | |
|---|---|---|---|
| | | Million acres | Percent U.S. total |
| **Original public domain** | | | |
| Cessions by original states | 1781-1802 | 237 | 10.2 |
| Louisiana Purchase | 1803 | 560 | 24.2 |
| Florida Purchase | 1819 | 46 | 2.0 |
| Oregon Compromise | 1846 | 183 | 8.0 |
| Mexican Treaty | 1848 | 339 | 14.6 |
| Purchase from Texas | 1850 | 79 | 3.4 |
| Gadsden Purchase | 1853 | 19 | .8 |
| Alaska | 1867 | 375 | 16.1 |
| Subtotal | | 1,838 | 79.3 |
| **Never public domain** | | | |
| Original states | | 305 | 13.2 |
| Texas | | 170 | 7.3 |
| Hawaii | | 4.1 | .2 |
| Subtotal | | 479.1 | 20.7 |
| **Total** | | 2,316.1 | 100 |

million acres—were opened to homesteaders. Through the mid-1930s, federal policy was to sell or give away the public lands to private owners and to states so that the nation would be "tamed, farmed, and developed" (Coggins and Wilkinson, 1987) (Figure 2-2). By 1976, more than 1 billion acres of the public domain had been disposed of under public land laws.

Public land laws usually distinguish between the public domain—lands obtained by the United States from other sovereigns—and acquired lands—lands obtained from private or state ownership by gift, purchase, exchange, or condemnation (Coggins, 1991). Today's federal lands consist of the remains of the original public domain in states west of the original 13 colonies, some of their claims, Vermont, and lands that have

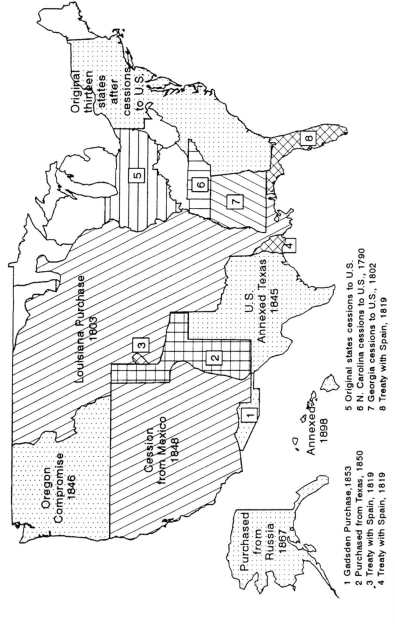

**FIGURE 2-1** Acquisition of U.S. territory from foreign nations.

# 38　SETTING PRIORITIES FOR LAND CONSERVATION

**FIGURE 2-2**　Indian lands, October 1991.

# OWNERSHIP AND MANAGEMENT 39

Source: Bureau of Indian Affairs, 1991.

been acquired by the federal government. The remains are public domain lands that were not claimed by settlers or miners under various homestead and other disposal laws or were not granted to states, railroads, and other entities to promote development and expansion of settlement.

**Designation of Federal Lands**

Starting with Yellowstone Park in 1872, some of the public domain was reserved in federal ownership as national parks, monuments, and forests. Formally, a reservation is a federal tract of land that Congress or the president has dedicated for particular uses (Coggins, 1991). The bulk of the reservations outside of those in Alaska were made from the 1870s through the 1930s. Federal reservations from the public domain in Alaska largely were created by the Alaska National Interest Lands Conservation Act of 1980 (ANILCA). The federal rangelands now managed by the Bureau of Land Management (BLM) effectively were closed to settlement in 1934, although the policy of retaining most of them in federal ownership was not made explicit until 1976 with passage of the Federal Land Policy Management Act (FLPMA).

Some shifts in status of and jurisdiction over the federal lands still occur. The parks and refuges created by ANILCA shifted jurisdiction mainly from BLM to the National Park Service (NPS) and the Fish and Wildlife Service (USFWS). Other federal lands have been shifted from the Forest Service (USFS) to NPS and from BLM to the other three federal agencies.

Lands reserved from the public domain decades ago and federal lands more recently added usually are not solid blocks or units. Lands reserved since 1872 for national forests, parks, and refuges generally were unsuited for private ownership, in large part because much of the land was semiarid. Homestead laws gave little recognition to the basic condition of the lands, but homesteaders generally chose watered land. Valleys and riparian corridors that provided access to water, natural vegetation, and opportunities for irrigated pastures were homesteaded in fingers reaching up streams, with higher, unwatered public land on both sides; many of the inholdings in the West today are the legacy of such homesteads.

Much of the public lands and some of the lands now managed by

BLM were overgrazed, and large portions of the public domain lands still are poorly suited for private ownership of large populations. Most of the national forests and BLM lands continue to be available for mineral prospecting and, under the provisions of the Mining Act of 1872, ultimately are available for transfer to private ownership if mineral deposits are discovered. Particularly in the case of NPS, the inholdings often are considered priority acquisition sites.

Nearly all of the national parks and grasslands and most of the national wildlife refuges and forests east of the Rockies are acquired lands. The national parks and wildlife refuges were acquired mainly for their particular natural-resource characteristics. The national forests and grasslands, however, often were acquired for other reasons; for example, abandoned farmlands were acquired during the Great Depression. The only large areas of federal lands reserved from the public domain east of the Great Plains are the Superior National Forest in Minnesota and the Ouachita National Forest in Arkansas and Oklahoma.

The 1911 Weeks Act, under which most of the eastern national forests were established, provided for land acquisition to protect the headwaters of navigable streams and to ensure continuous supplies of timber. The lands were in poor condition—cutover and burned—from the wave of logging that took place from 1850-1920; those conditions and burdensome property taxes made economic management as private timberlands infeasible.

The federal government acquired rundown farmlands under the federal emergency land utilization (LU) program from 1935 to 1937 and under the Bankhead-Jones Farm Tenant Act from 1938 to 1946. The responsibility for restoring the lands was given to the Soil Conservation Service. Of the roughly 11.3 million acres of LU lands, about 2.2 million acres ultimately were transferred to BLM, 1.5 million acres were incorporated into eastern national forests, and 3.8 million acres in the Great Plains formed the national grasslands, which are administered by the USFS (Wooten, 1965). The remaining lands were conveyed to various other parties.

## Federal Landownership

Public land comprises the holdings of the federal government, as well as those of state and local governments (Indian lands held in trust are discussed in Chapter 3). All three land accounts have grown as a result of LWCF outlays. State and local land holdings compose 8% of the total U.S. land base exclusive of Alaska and are increasing (Wolf, 1981). Like federal lands, state and local holdings are concentrated in the West and are a small proportion of total land in the nation, ranging from 2% in the South to 11% in the West. In a few states, such as New York, state and local holdings exceed federal holdings. Most government reports estimate that federal lands constitute roughly one-third of the nation's land base. If only the West is considered, as much as 48% of the land is federally owned; 63% of the West is federal land when Alaska is considered.

Federal landownership in the United States has fluctuated greatly with time. Originally, this was because of federal acquisitions and annexations; more recently it follows from federal commitment to protect land for aesthetic, recreational and environmental reasons. Whereas in the nineteenth century public lands were sold to mobilize private development, in the present century there has been an impetus to buy back federal lands. From the 1890s through the 1930s, federal land policy in America gradually shifted from disposition toward reservation of the remaining public domain. By the middle of the century, federal ownership stabilized briefly at 760 million acres, or 33% of the total U.S. land base of 2.275 billion acres. This includes 365 million acres in Alaska, nearly all of which was federal land before statehood in 1959.[1]

Despite heightened environmental awareness in the 1970s, concerns about further federal land acquisition persisted. In 1979, the General Accounting Office stated that current federal land policies tended to overlook least expensive land protection strategies, adversely affected private landowners, and underestimated facility maintenance costs of the expanding land base (GAO, 1979). Early in the 1980s, the Reagan

---

[1] At statehood, Alaska was granted slightly more than 104 million acres from federal lands. Alaska made its selections gradually as federal lands were surveyed. Furthermore, the settlement of native claims in Alaska under the Alaska Native Claims Settlement Act of 1971 led to the transfer of about 44 million acres of federal land to native villages and regional corporations. Thus, the total acreage of federal land in Alaska decreased considerably during the past 3 decades. In the rest of the country, however, the federal land base grew slowly.

administration made known its intentions to sell as much 5% (35,000,000 acres) of the federal lands for budget relief (Short, 1989) and to place a moratorium on LWCF appropriations. The latter pledge may account for the downward trend in LWCF appropriations (Figure 2-3). Federal acreage purchased was abnormally low in the early years of the Reagan administration (Table 2-2).

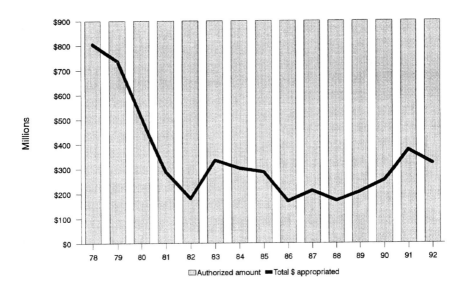

**FIGURE 2-3** Land and Water Conservation Fund, fiscal years 1978-92

The final years of the Reagan administration were marked by an executive order creating the President's Commission on Americans Outdoors (PCAO). The PCAO's mandate in some ways resembled that of ORRRC, but it made no assumption that additional public lands were necessary to achieve the public health and vitality associated with out-

TABLE 2-2  Federal Acreage Purchased Using Land and Water Conservation Fund Money

| Year | Forest Service | National Park Service | Fish & Wildlife Service | BLM* | Total |
|---|---|---|---|---|---|
| 1965 | n/a | 729 | n/a | n/a | 729 |
| 1966 | 29,437 | 3,974 | n/a | n/a | 33,411 |
| 1967 | 70,058 | 52,671 | 3 | n/a | 122,732 |
| 1968 | 85,861 | 58,522 | 2,561 | n/a | 146,944 |
| 1969 | 91,327 | 98,355 | 832 | n/a | 190,514 |
| 1970 | 79,720 | 70,540 | 15,031 | 0 | 165,291 |
| 1971 | 80,944 | 44,651 | 4,530 | 536 | 130,661 |
| 1972 | 75,469 | 68,499 | 11,506 | 594 | 156,068 |
| 1973 | 91,138 | 49,643 | 3,017 | 2,117 | 145,915 |
| 1974 | 35,933 | 86,277 | 2,859 | 4,599 | 129,668 |
| 1975 | 44,051 | 75,412 | 2,625 | 1,744 | 123,832 |
| 1976 | 57,235 | 149,246 | 25,064 | 1,903 | 233,448 |
| 1977 | 39,472 | 218,245 | 27,405 | 1,840 | 286,962 |
| 1978 | 32,429 | 260,555 | 26,599 | 5,249 | 324,832 |
| 1979 | 94,861 | 122,167 | 37,842 | 3,400 | 258,270 |
| 1980 | 66,049 | 53,702 | 21,540 | 3,760 | 145,051 |
| 1981 | 102,463 | 18,185 | 21,202 | 2,119 | 143,969 |
| 1982 | 8,332 | 18,061 | 14,101 | 185 | 40,679 |
| 1983 | 10,447 | 13,285 | 18,921 | 12,049 | 54,702 |
| 1984 | 19,162 | 14,369 | 164,448 | 2,004 | 199,983 |
| 1985 | 34,448 | 27,051 | 79,620 | 151,613 | 292,732 |
| 1986 | 42,860 | 73,689 | 28,663 | 72,842 | 218,054 |
| 1987 | 85,832 | 17,645 | 55,268 | 48,161 | 206,906 |
| 1988 | 57,768 | 43,577 | 88,543 | 57,465 | 247,353 |
| 1989 | 106,354 | 29,029 | 83,216 | 72,475 | 291,074 |
| Total | 1,441,650 | 1,668,079 | 735,396 | 444,655 | 4,289,780 |

Source: The Conservation Fund.
*Includes land acquired by exchange.

door recreation.[2] As with ORRRC, the PCAO noted that most of the nation's population was in the East, whereas most of the federal lands were in the West, a regional equity issue requiring attention. Perhaps more important, the PCAO substantiated ORRRC's earlier estimates of recreational demand. Indeed, some levels of outdoor use and participation projected by ORRRC for the year 2000 were surpassed as early as the 1970s (PCAO, 1986). Figure 2-4 provides an overview of user demand for NPS facilities, a partial indicator of this demand.

## Private Landownership in the United States

Alexis de Tocqueville, traveling in the United States early in the nineteenth century, made note of the unalloyed devotion to private ownership in the new republic:

*In no country in the world is the love of property more active and more anxious than in the United States; nowhere does the majority display less inclination for those principles which threaten to alter, in whatever manner, the laws of property* (cited in Sakolski, 1957).

Of the 2.27 billion acres composing the United States today, 1.35 billion are owned privately. Although a majority of the American land base is privately owned, the 1980 census data (the most recent government information available), show that the nation's private real estate was held by only 34 million owners, or approximately 15% of the population (Lewis, 1980). The distribution of holdings among the owners is also skewed: 5% of landowners (including corporations) have title to 40% of private lands, and the top 5% own three-fourths of the total. In contrast, the bottom 78% of those who own land own 3%. Private landownership is most concentrated in the West and least concentrated in the north central states (Gustafson, 1983).

---

[2]PCAO did, however, argue for a dedicated trust fund of approximately $1 billion per year for recreational ends (Madson, 1988), a fund many have interpreted as an extension of the LWCF. LWCF is not strictly a trust fund, because authorizations that are not appropriated return to the federal government for discretionary use.

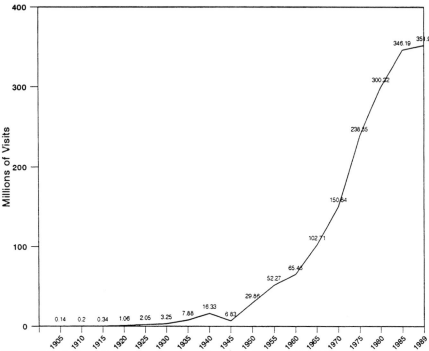

**FIGURE 2-4** Visits to national parks, in millions, 1905-89. Source: NPS, 1990.

Private landownership has changed over time. As the land base of the nation increased, the federal government's share at times grew to 80% of the total. With the help of numerous homestead acts, cash sales, and grants to railroads, private landownership expanded to well over half of the nation at some times (Clawson, 1973). Despite the central place of private ownership in American life and culture, however, little systematic federal information has been kept on who owns what land. With the exception of the Agricultural Census of the Department of Agriculture, which covers only agricultural lands, the federal government publishes only piecemeal information on private ownership (Geisler, 1983; Gustafson, 1983).[3]

---

[3]Government time-series data on private forest landownership are presented in acres rather than numbers of owners (e.g., Waddell et al., 1987), preventing trend analyses of ownership.

Available data confirm that a small fraction of private landowners control most of the land in private hands, and this concentration does not seem to be abating. The proportion of privately owned lands and the distribution of ownership among a few are of considerable interest in light of the consistently high percentages of Americans supporting expanded federal ownership in public opinion surveys (see Appendix C). Public ownership might be viewed as a way to access land in the absence of widespread opportunity or material means to own it. On the other hand, periodic public clamor against federal ownership as a barrier to private ownership might be more directed protest over private land concentrations and the ownership access obstacles it presents.

## DISINCENTIVES FOR CONSERVATION

Many incentives and disincentives affect the ownership, management, and use of federal and other public lands. Some of those are the lack of market pricing for some resources, accrual of costs or benefits to some who are not direct parties to transactions, uneven effects of government interventions, and structural problems in particular markets (e.g., long periods required to renew resources). For example, in the market economy, near-term results are given priority over long-term results, which can encourage farming practices that lead to soil erosion and rapid cutting of timber. In addition, high inheritance taxes can lead to fragmentation of large land holdings—which also favors short-term over long-term results. And farmers have long been paid for draining wetlands and planting croplands from roadside to roadside, although payment for the former practice was halted with the passage of the 1985 farm bill.

Private owners have little incentive to manage lands for conservation of biological diversity or other values for which markets do not exist, at least in the short term. Yet without the cooperation of private landholders, the size of individual holdings might limit the conservation actions of landowners, because large areas often are necessary to maintain biological diversity and avoid fragmentation of ecosystems.

## THE ROLE OF LANDOWNERSHIP IN CONSERVATION

Public landownership has been an important aspect of America's social and economic history. How much of the nation should remain in federal ownership is a contentious issue fueled by political philosophies, changing perceptions of the public interest, and different impressions about private ownership trends and the "American Dread." But in the long run, the wisdom of expanded federal ownership in America's future is contingent on historical circumstances well beyond a scoreboard summary of public- versus private-sector ownership trends. Factors worth pondering include population growth and distribution, the nation's changing occupational structure, new institutional developments that might lessen the traditional polarity between public and private landowner, and growing sophistication about conservation at the ecosystem level.

The potential effect of population change on public ownership is striking. By 2010, the nation's population is expected to increase by 44 million people (PCAO, 1986)—people who are expected to have an increasing appetite for outdoor recreation in many forms. As population grows, the national land base shrinks relatively (per capita), and less land is available for each new generation. Another relevant demographic characteristic is age structure. Overall, the American population is aging, and recreational and aesthetic uses of the public land are changing as a consequence.

In ways that often go unrecognized, the American land base is also shrinking in an absolute sense. This occurs as public and private lands undergo changes that eliminate certain uses, e.g., open space used for recreational or residential sites. Such changes include designation of lands as Superfund and hazardous waste sites or lands subjected to soil erosion and aquifer depletion, overgrazing, atmospheric pollution, and conversion to highways. Land dedicated for highways already is roughly equivalent to lands held by NPS (Klockenbrink, 1991). And 13 million acres of U.S. rangeland show signs of severe desertification (Dregne, 1991). Relative and absolute shrinkage in the nation's land base warrant serious consideration in planning and land allocation to public- and private-sector uses.

America has undergone and faces further occupational restructuring, with an associated redefinition of land-use needs and ethics. As service-sector employment grows at the expense of manufacturing and more basic extractive livelihoods, land assumes new importance as a consumptive, recreational, and aesthetic good, reducing further a direct survival relationship between the population and the land. As use requirements change, so will ownerhsip imperatives and the forms of public and private ownership.

New forms of public-private ownership dull the distinctions between public and private. Those new forms include land trusts and intergenerational lease-back mechanisms, easements and less-than-fee management options, covenants between private owners, development rights that are purchased or transferred, compensatory zoning, reserved and dedicated rights, and other evolving mechanisms that partition equity in land to benefit public and private interests. Such quasi-public institutions are proliferating and offer alternatives to the proponents of traditional public or private ownership and to planners who must try to balance the interests of each.

The committee's study concluded that public and private values cannot be conveniently separated. The vigorous pursuit of public values no longer takes place only on public lands or out-of-the-way preserves and set-asides. Just as federal lands host a broad array of private uses and ownership rights, private lands are shouldering an increasing public responsibility in the areas of conservation, environmental protection, and public-interest health and recreation.

# 3

# The Land-Acquisition Process

The process of federal land acquisition involves interactions among a variety of participants. Some of the participants—federal agencies, Congress, local governments, and landowners—have official decision-making or administrative authority; others—such as owners of adjoining property, and national and local interest groups—lack formal authority but might have considerable influence on those who do. This chapter reviews the mandates of the federal agencies that acquire land for conservation purposes and the sources of funding available for acquisition. The chapter then explores who exercises authority and influence in the acquisition process and their powers, responsibilities, and modes of interaction, and analyzes how the process actually works to balance often competing interests. Specific consideration is given to how the process incorporates scientific information relevant to reaching conservation objectives and how it accounts for the interests of various groups.

## SOURCES OF FUNDING

Until 1964, each agency had a distinctive funding source for acquisitions. The Forest Service (USFS) drew a modest annual appropriation from the Weeks Act, and the Fish and Wildlife Service (USFWS) used monies accumulated in the Migratory Bird Conservation Fund. The infrequent National Park Service (NPS) acquisitions were funded by special congressional appropriation. The Bureau of Land Managment (BLM) had little involvement in land aquisition.

### The Land and Water Conservation Fund

The availability of funds for federal land acquisition increased dramatically with the passage of the Land and Water Conservation Fund Act of 1964. The Land and Water Conservation Fund (LWCF) is a special account in the U.S. Treasury from which Congress annually appropriates money to acquire lands for conservation and recreation by federal and state agencies. Certain federal revenues, including the proceeds of surplus federal property sales, the federal motorboat fuels tax, and a portion of Outer Continental Shelf leasing receipts, are credited to the LWCF to provide it with a maximum of $900 million annually. None of this money can be spent, however, unless specifically appropriated by Congress.

Since the act's passage, more than $3.2 billion of matching grants from the LWCF have been made to states to enable them to plan, acquire, or develop qualifying projects. In the same period, more than $3.6 billion has been expended from the LWCF for land acquisition by the Department of the Interior and the Department of Agriculture. Appropriations from the LWCF have varied considerably from year to year, from nearly $800 million in 1978 to less than $200 million in 1982. In 1991, LWCF appropriations were $374,943,000. The act specifies that not more than 60% of the money appropriated from the LWCF each year is to be for grants to the states, and the state share has typically been much less than that amount; in 1982, no LWCF monies were appropriated for state grants.

The idea for the LWCF can be traced to a proposal by Stewart Udall in 1961 for a fund for federal land acquisition and to the Outdoor Recreation Resources Review Commission (ORRRC), which was established by Congress in 1958. When its final report was released in 1962, the ORRRC's overall conclusion was that the demand for outdoor recreation had grown dramatically since World War II and was likely to continue to do so, necessitating a major governmental effort to provide the land for such purposes. The ORRRC also noted that although much of the available public recreation land was in the West, much of the nation's population was in the East. The congressional committees that were responsible for the legislation creating the LWCF agreed that "a substantial part" of the monies to be devoted to federal acquisitions should go toward the purchase of privately owned inholdings within the authorized boundaries of national parks, forests, and refuges (U.S. Congress, House, 1963; U.S. Congress, Senate, 1964). Another aim was to establish publicly significant recreation areas within easy distance of major population centers.

### National Park Service

Throughout its history, the national park system as administered by NPS has been a principal beneficiary of the LWCF. Primarily because of backlogs in the acquisition of inholdings within the park system, Congress has on several occasions increased the authorized ceiling for the LWCF to its present level of $900 million (Glicksman and Coggins, 1984).

### U.S. Forest Service

The USFS is another federal beneficiary. Although the 1897 law creating the USFS made no mention of wildlife or public recreation, those have been among the purposes for which national forests have been managed since the agency began operation. Passage of the Multiple Use and Sustained Yield Act of 1960 heightened the importance of those purposes. The national forest system as a whole is to be managed for multiple uses, but the management of particular areas is likely to emphasize only one or a few uses. That is reflected in the limited authority conferred by the Land and Water Conservation Fund Act for land acquisition using monies from the LWCF: The only areas that may be acquired by the USFS using LWCF monies are inholdings within national forest wilderness areas, inholdings within other national forest areas that "are primarily of value for outdoor recreation areas," and areas not to exceed 3,000 acres that are adjacent to an existing national forest boundary and that would compose "an integral part of a forest recreational management area."

## U.S. Fish and Wildlife Service

The third federal agency for which the LWCF can be tapped for land acquisition is the USFWS. The Migratory Bird Conservation Act (MBCA) in 1929 gave USFWS (then known as the Bureau of Biological Surveys) the authority to acquire lands for inviolate sanctuaries for migratory birds. The enactment of the Migratory Bird Hunting Stamp Act 5 years later provided a special mechanism for funding such acquisitions. The Fish and Wildlife Act of 1956 conferred a general grant of authority upon the USFWS to acquire refuge lands without regard to the inviolate sanctuary provisions of the MBCA and without necessarily being limited to lands of value to migratory birds.

The purposes for which lands can be acquired by USFWS using LWCF monies have expanded steadily. Until 1962, the question of whether public recreation was an intended purpose of national wildlife refuges had not been addressed legislatively, although the MBCA since 1929 had empowered USFWS to permit hunting of migratory birds on its refuges. In 1962, Congress passed the Refuge Recreation Act. That act authorized the secretary of the interior to administer national wildlife refuges or parts thereof for recreation and "an appropriate incidental or secondary use" if it was determined that such use was compatible with the primary purposes for which those areas were established. It also authorized the secretary to acquire lands for recreational development adjacent to wildlife refuges, but stipulated that monies in the Migratory Bird Conservation Fund could not be used for that purpose. At the time, however, no other sources of funding were available for acquisitions. Two years later, when Congress enacted the Land and Water Conservation Fund Act, it authorized USFWS to use LWCF monies to acquire lands for the incidental recreation purposes of the Refuge Recreation Act.

In 1973, USFWS's land-acquisition authority was expanded significantly when Congress enacted the Endangered Species Act (ESA). That act authorized the secretary of the interior to acquire lands needed for the conservation of endangered or threatened species. It specifically authorized the use of the LWCF for such purposes and amended the Land and Water Conservation Fund Act to reflect that authorization. This amendment was the first authorized use of the LWCF that was not tied explicitly or implicitly to outdoor recreation resources. A similar amendment followed in 1976, when the Land and Water Conservation

Fund Act was amended to allow USFWS to tap the LWCF for land acquisition of any refuge area authorized by specific act of Congress, as well as refuges to be acquired under the general grant of land-acquisition authority found in the Fish and Wildlife Act of 1956. The only limitation was that the LWCF could not be used to acquire lands authorized for acquisition under the MBCA.

In 1986, the LWCF Act was amended again. The Emergency Wetlands Resources Act directed the secretary of the interior to establish a plan that specifies the wetlands that should be given priority for federal or state acquisition. In putting together this plan, the secretary is to take into account the value of particular types of wetlands for certain purposes, among them wildlife (including threatened or endangered species) and outdoor recreation. The secretary is authorized to acquire wetlands that are not acquired under the authority of the MBCA, consistent with the wetlands conservation plan. The Emergency Wetlands Resources Act also amended the Land and Water Conservation Fund Act to allow the use of the LWCF to acquire priority wetlands.

### Bureau of Land Management

The fourth significant recipient of LWCF monies is BLM. The Federal Land Policy and Management Act of 1976 (FLPMA) authorizes land acquisition by BLM but does not specify the source of funds for such acquisition. The Land and Water Conservation Fund Act does not make explicit mention of BLM, but LWCF monies have been used to acquire land for outdoor recreation by BLM since at least the early 1970s. Authority for the secretary of the interior to acquire land for endangered or threatened species applies to the national wildlife refuge system, which is administered by USFWS however, the ESA authorizes land acquisition by the secretary without specifying whether it may be done only through USFWS or other Department of the Interior agencies, such as BLM.

## ACQUISITION BY FEDERAL AGENCIES

Each federal agency acquires lands in pursuit of its own legislative mandate. Acquisition priorities and strategies are also influenced by longstanding agency practices. The overriding factor in land acquisition is funding: The administration must identify priorities among those identified by the four agencies (see Appendix B).

The enthusiasm of agencies is tempered by the difficulty in acquiring lands and the cost of managing lands after acquisition. At minimum, acquiring new lands entails expenditures for boundary maintenance,[1] protection of public safety, and payments to local governments in lieu of taxes that would have been received if the lands were in private ownership. Appropriation by Congress of money for acquisition does not necessarily mean that Congress will appropriate funds for management or authorize staff positions. New federal lands sometimes also entail expanded responsibilities for federal agencies. For example, the Santa Monica Mountains National Recreation Area gave NPS substantial new responsibilities for providing urban recreation in the Los Angeles region, and the use of the greenline park concept has led to significant demands for land-use planning and coordination with state and local governments. Expanded responsibilities may be costly and can present administrative challenges for which existing agency staff are unprepared.

A relatively new complication affecting acquisition is the problem of toxic contamination, which can permanently bar acquisition—a parcel of land that poses a health or safety threat due to the presence of toxic substances or other hazards does not meet administration criteria for acquisition.

Acquisitions along the New River in West Virginia illustrate the problems created by the presence of toxic waste. Some parcels within the New River National River area (administered by NPS) are privately owned. One parcel includes extensive river shoreline and was for sale in 1992; a coal mine waste dump is part of the parcel. NPS would have liked to purchase the parcel; however, it is precluded from purchasing the waste dump, and the seller was unwilling to sell less than the whole parcel.

Land-acquisition authority includes purchase of land in fee, with or without condemnation by eminent domain (i.e., through sales, exchanges, gifts, bequests, and other means), as well as the acquisition of lesser interests (easements, rights of way, life estates, etc.). The four federal

---

[1] However, acquiring inholdings and adjacent parcels to consolidate existing holdings sometimes reduces the length of boundaries and, therefore, lowers associated costs.

agencies examined by the committee have the authority to take land from unwilling sellers by virtue of eminent domain. That authority often appears in implied, redundant, and qualified form. Public land laws are well known for complexity, ambiguity, and duplication.

## The National Park Service

Congress created NPS in 1916, nearly 50 years after the first national park was established at Yellowstone. The U.S. national park system was an innovation in land use (Runte, 1990), and after its creation, countries throughout the world used it as a model in establishing similar systems.

The national park system has 361 units consisting of approximately 80 million acres, of which 76 million are federally owned (Figure 3-1). Sixty-eight million originally were in the public domain, and most were withdrawn by federal statute, not always with NPS participation and support. Fifty-four million acres are in Alaska. Two-thirds of the units include land not owned by the federal government; some of the oldest units in the national park system, including Yosemite and Yellowstone, have inholdings. One of the newest units, Great Basin, has no inholdings. NPS estimates it would cost more than $1 billion in today's dollars, perhaps as much as $2 billion, to purchase all inholdings. In the past few years, the LWCF appropriation for NPS has averaged from $50-60 million.

### Mandates

From the reservation of Yosemite Valley as a state park in 1864 to the National Park Service Organic Act in 1916 and through subsequent political battles, several social philosophies and economic interests coalesced to form the foundation of a system of national parks (Runte, 1990; NRC, 1992a). Park supporters have included preservationists wanting to keep nature unspoiled, progressive conservationists concerned that average citizens have easy access to the nation's most spectacular scenery, scientists interested in preserving areas for research, and business people promoting tourism or protecting future raw material sources.

The diverse concerns of park supporters were reflected in the con-

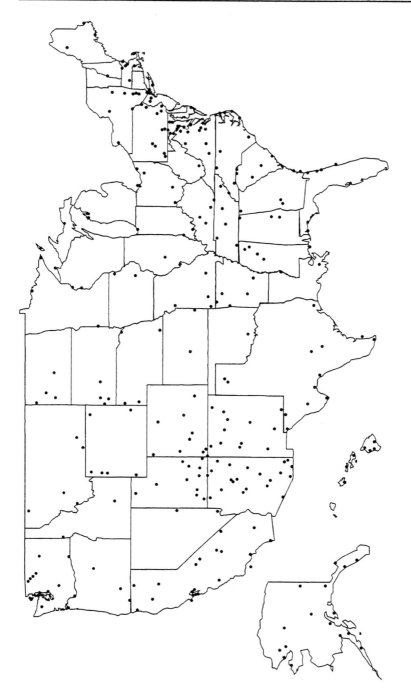

FIGURE 3-1  The national park system.

gressional mandate to the agency. NPS's assignment was to "conserve the scenery and the natural and historic objects and the wild life therein and to provide for the enjoyment of same in such manner and by such means as will leave them unimpaired for the enjoyment of future generations" (16 USC 1).

The dual objectives of preservation and use have been the source of numerous controversies over national park management. Proponents of various park visions often disagree on how best to protect park resources while providing for recreational use.

The national park system now comprises numerous land categories, including national parks, monuments, historic sites, battlefields, and recreation areas. Each individual park unit is created by an act of Congress. The authorizing legislation generally confirms that the unit is to be managed according to general rules governing the system and defines management goals for the particular unit. Some units' authorizing legislation is very specific; others have ambiguous language.

Overall conservation objectives expressed in management policies have changed with increased ecological understanding and use demands. Objectives have included prevention of poaching; control of elements of the ecosystem considered undesirable—e.g., at first fire and predators, then human activities that prevent fire and predator control; human-use management to prevent resource degradation; exploration of ways to mitigate effects of human activities on parks as well as land outside of parks; and integration of parks into larger, regional land-use management patterns to sustain regional biological diversity. Overall NPS policy is to maintain natural processes responsible for the continuing evolution of natural ecosystems and to restore elements lost as a result of human activities (Keystone Center, 1991).

## Acquisition

Acquisition of private lands did not become an important concern of NPS park managers until the 1960s. Before then, most new parks were created from the public domain or from national forests, or in a few cases, by donation. In 1961, Congress established the Cape Cod National Seashore and authorized federal money for parkland acquisition from private owners; the 1964 Land and Water Conservation Fund Act provided for future purchases. Congress has been active in authorizing additions to the system but the LWCF never has been adequate to meet these authorizations. NPS has had a persistent backlog of congressionally authorized but unpurchased lands.

Today, acquisitions are of two types (NPS, 1988). For parks authorized before 1959, acquisition is made as opportunity is presented; those units generally have little private land, and unless development poses a threat to park resources or owners wish to sell, federal acquisition is not urgent. Parks authorized after 1959 often include considerable inholdings, and the acquisition program is more systematic.

Some important characteristics of NPS are revealed in its response to two changes that have created additional opportunities for land acquisition during the past several decades. The first was the addition of urban parks, such as the Golden Gate National Recreation Area in California and the Fire Island National Seashore in New York. Urban parks created a new set of management challenges for park rangers. During the 1960s, Secretary of the Interior Udall emphasized the importance of bringing conservation to the cities and improving the quality of urban life. These goals were consistent with the emerging environmental movement and President Johnson's War on Poverty. But the biophysical and social contexts of urban park management were new to NPS, and although the urban parks created a new avenue for system expansion, they also created new problems for the agency (Foresta, 1984).

The second change was the increasing emphasis on biological diversity and ecosystem management in the environmental and scientific communities. Those concerns have been articulated along with demands for large nature reserves or wilderness areas. Historically, NPS has seen its mission as managing recreation rather than resources. Thus, demands for "big wilderness," while presenting an additional opportunity for expansion, also challenge traditions within the agency.

**Acquisition Criteria**

The NPS organic act has no acquisition authority, although a variety of laws have filled the void. By one count, 41 different pieces of legislation approve some version of eminent domain to be exercised by the secretary of the interior (Hemmet, 1986). Sometimes, the authorizing legislation for individual park units limits methods of acquisition, for example, to donations or exchanges, to maximum acreage, or to less-than-fee interest. The Ebey's Landing National Historic Reserve in Washington, for example, was established by Congress in November 1978 "to preserve and protect a rural community which provides an unbroken historical record from the nineteenth-century exploration and settlement in Puget Sound to the present time" (National Parks and Recreation Act of 1978, 16 U.S.C.A. Section 461). The authority for the acquisition of the property forbids the use of eminent domain. The reserve is made up of "a scenic island community of farms, woodlands, open space, historical structures, and the historic town of Coupeville. The resources to be protected constitute the historic rural environment of central Whidbey Island," and cover 13,100 acres of land and 4,300 surface acres of salt water.

Legislative acquisition strategies also can encourage local zoning. An example of federally encouraged local zoning is found in the Cape Cod National Seashore Act of 1961 (16 U.S.C.A. Section 459b-1), which gives the secretary of the interior the power "to acquire [for the national seashore] by purchase, gift, condemnation, transfer from any Federal agency, exchange or otherwise, the land, waters, and other property, and improvements thereon." But the condemnation powers are constrained in several ways. Owners whose land is condemned may elect to retain the right of use and occupancy of the property for residential purposes for as long as 25 years. The act also suspends the secretary's authority to acquire improved property (e.g., single-family dwellings) by condemnation if the towns within the seashore have valid zoning laws approved by the secretary and applicable to the property. The Cape Cod Seashore has adopted use guidelines for private property that direct private owners of improved property to comply with the act; this formula effectively created federal zoning in the form of indirect federal control over local land-use decisions.

Limitations on acquisition authority and criteria for priorities within individual units are reflected in land-protection plans prepared for each NPS unit. Key considerations for establishing land-acquisition criteria are the primary purpose of the park, land price escalation, legislative history, imminent threats, and protection of the park. Current administration policy requires land-protection plans to identify, for each privately owned parcel, the least federal interest necessary to achieve the goals of the park. Land-protection plans identify nonfederal tracts, identify the interest needed in each tract, and establish priorities. Acquisition authority under some enactments (such as the National Trails System Act) is limited to a certain number of acres per mile in fee simple. And under the W & SRA, once 50% of any type of unit is in public ownership, NPS can no longer condemn an easement.

NPS developed ranking criteria for parcels in different units during the 1980s. Under those criteria, projects were ranked according to regional acquisition priorities. Considerations included

- The type of area;
- Whether legislation was needed;
- Whether plans were completed;
- Whether there was a congressional or executive mandate;
- Location, number of tracts and acres;
- Cost;
- Whether it was key to accomplishing a mission defined in plans (e.g., to provide access to larger tract of public land, or protect key natural or cultural features);
- Probability of damage within 3 years and permanence of damage;
- Whether it protected an established area;
- Population within 1 and 2 hours of driving time;
- Availability of acquisition alternatives;
- Operation, maintenance, and development costs;
- Development and timing;
- Willingness of seller;
- Whether condemnation authority existed and willingness to use it;
- Organizational capability;
- Local support;
- Whether congressional oversight or approval was required;
- Whether it was coordinated with other planning processes;
- Whether it would have completed or continued an existing project or started a new area;
- Whether the authorization was general or specific;
- Whether it was eligible for funding from a special account that was available under the 1978 Omnibus Parks Act of 1978.

These criteria never were fully implemented. In the early months of the Bush administration, the Federal Land Acquisition Priority Procedure was put into effect by OMB to rank the priorities of NPS, BLM, USFS, and USFWS.

NPS never intended to buy everything inside park boundaries. Priorities do change, e.g., because of owner hardship or threats to resources. Some boundary changes have been made to reduce the priority of tracts that were not necessary or already were developed, but there is no systematic understanding of the properties in the acquisition backlog.

According to NPS, most landowners are willing to sell for the right price. In condemnation proceedings, an independent third party establishes the price, but condemnation cases often are settled before going to court or by willing sellers. For the Appalachian Trail, for example, NPS negotiated with some owners 30 or 40 times before using condemnation authority. On other occasions, such as the Cuyahoga Valley National Recreation Area in Ohio, NPS has undertaken removal of people from farmlands and towns, many of whom were unwilling sellers.

## U.S. Forest Service

USFS has three general missions: management of the national forests; cooperation with the states in the protection of forests against wildfires, insects, and disease and in providing technical and financial assistance to private and other nonfederal forest owners; and forest-related research. Land acquisition mainly is related to the national forests, but the cooperative USFS protection and management programs for state and private forests might be relevant tangentially to national forest land acquisitions.

### Mandates

The 1897 Forest Service organic act provided that national forests be established only "to improve and protect the forest within the boundaries, or for the purpose of securing favorable conditions of water flows, and to furnish a continuous supply of timber for the use and necessities of the citizens of the United States." The clause "to improve and protect" was interpreted broadly. Uses other than those specifically mentioned in the 1897 act, including livestock grazing and recreation, were allowed. The Multiple-Use and Sustained-Yield Act of 1960 provided specific authority for five categories of use—outdoor recreation, range, timber, watershed, and wildlife and fish—and confirmed by statute what had been administrative policy for more than 50 years. Multiple-use was defined to include the "harmonious and coordinated management of the various resources, each with the other, without impairment of the productivity of the land"; the land remained open to mining, except where specifically withdrawn from application of the mining laws.

The national forests are described aptly as lands of many uses, with no one of the listed uses having automatic statutory priority over the others. Setting priorities locally is left to the land managers, and this policy was given additional statutory blessing in the land-use planning provisions of the National Forest Management Act of 1976. Although USFS had much earlier administratively set aside extensive areas of the national forests as wilderness and primitive areas, it was not until the Wilderness Act of 1964 that wilderness was added to the list of uses recognized in law. The authorized uses that can be pursued on wilderness areas designated by Congress are limited by provisions of the Wilderness Act. Since 1964, various other designations also have limited the uses that can be made of specific parts of the national forests. Designating areas of the national forests as national wild and scenic rivers, national trails, national monuments, scenic areas, and volcanic areas reduces the total area of the national forests available for multiple-use management, although it ensures a broader range of uses overall.

## Acquisition authority

Land-acquisition authorities for the national forests initially were broad and supported the overall missions for the national forests. The 1911 Weeks Act provided for the purchase of "such forested, cutover, or denuded lands within the watersheds of navigable streams as in [the secretary's] judgment may be necessary to the regulation of the flow of navigable streams or for the production of timber." Lands were purchased under the Weeks Act to support the broad missions for the national forests, but each acquisition had to be based on regulating stream flows or producing timber.

The Land and Water Conservation Fund Act provided for land acquisition for the national forests for inholdings within wilderness areas and other areas primarily of value for outdoor recreation purposes. The latter limitation was an attempt to keep USFS from using the act as a vehicle for acquisitions for general forest management or expansion purposes (F. Gregg, USFS, pers. comm., June 21, 1990).

Various acts that designate federal lands for specific purposes add to USFS land-acquisition authority. One compilation identifies 18 significant statutory fragments that control the agency's acquisition practices (Lewis, 1978). The National Wild and Scenic Rivers Act, for example, provides land-acquisition authority to meet the purpose of the act, which is to protect rivers and their immediate environments for the benefit and enjoyment of present and future generations. The act authorizes use of LWCF appropriations. The National Trails Act provides authority to acquire lands to meet the purposes of the act within the boundaries of federal areas, as well as outside of federal areas if state or local governments fail to acquire land or enter into satisfactory agreements with landowners. The broad purpose of that act is "to provide for the ever-increasing outdoor recreation needs of an expanding population . . . (i) primarily, near the urban areas of the Nation, and (ii) secondarily, within scenic areas and along historic travel routes of the Nation, which are often more remotely located." FLPMA authorizes USFS land acquisition to provide access to national forests over nonfederal lands. Other acts establishing specific units, such as national recreation areas, provide additional land-acquisition authority.

Because of the numerous statutory provisions, criteria for setting USFS acquisition priorities must serve a panoply of purposes and uses, some of which give greater statutory priority in land acquisition than others (e.g., wilderness or wild and scenic rivers), despite the evenhanded treatment associated with the Multiple-Use and Sustained-Yield Act.

Regulations for the national forests provide for land-use plans for each national forest; each plan is accompanied by an environmental impact statement and is revised on a 10- to 15-year cycle. The planning regulations do not mention land acquisition as one of the matters to be discussed in the plans, but because the plans concern the overall missions and mandates for the national forests, many plans do address land-acquisition goals. Those activities are required to be consistent with forest and resource management plans.

The current planning regulations detail the factors to be considered during the planning process. In addition to goals for national forest outputs (e.g., timber and grazing), the regulations include goals such as maintaining viable populations of native and desired nonnative species well distributed throughout their geographic ranges and protecting and restoring natural biological communities; conservation of biological diversity, including recovery of threatened and endangered species; sustaining population viability of species that are sensitive to anticipated trends in environmental conditions or human activities; protecting rare, unique, and highly productive communities of plants and animals; and managing habitats and populations to produce ecological conditions that sustain human uses of species desired as commercial, recreational, or subsistence resources.

**Acquisition Criteria**

Land-acquisition needs initially are identified in individual forest plans. Some are very general and say, in effect, that lands will be acquired as they become available for purchase and meet the plan and USFS priorities. For example, the recent plan for the Green Mountain National Forest in Vermont states that highest priority for acquisition "will be given to tracts which: are near the Appalachian and Long Trails; are within or adjoining Wildernesses and Management Area 6.1 [a land-use category], where Primitive recreation is emphasized; have uncommon or outstanding qualities which make them special; adjoin significant streams; have important wildlife habitats; or consolidate public ownership" (USFS, 1989). The plan does not identify specific tracts that fit these priorities.

Plans for some other national forests are more specific in identifying priorities for land acquisition. For example, the 1987 plan for the Kootenai National Forest in northwestern Montana identifies 90,999 acres within the boundaries of the forest that are desirable for acquisition. It also identifies 68,922 acres that are desirable for disposal because they are isolated parcels, do not have the character associated with national forests, have management problems, or would contribute greater public benefits if they were state-owned. Areas to be acquired or that are available for disposal are identified on maps.

Reasons for acquisition of areas specified in plans cover a spectrum of concerns, including recreation, consolidation of national forest land, water frontage, isolated private parcels, big-game habitat, threatened or endangered species habitat, public access, improvement of timber production, essential fish or bird habitat, improvement of public management or use, protection of cultural resources, elimination of title problems, and cost-effective management (USFS, 1987a). The Kootenai National Forest plan identified areas for acquisition that would support the broad USFS missions. But many areas identified for acquisition in individual forest plans might not be considered of high priority once they are ranked against those from other national forests.

USFS uses a point system similar to the one used by the Office of Management and Budget (OMB) (see Appendix B) to rate properties and assemble information from the regions. Projects first must meet four minimum criteria, unless a project is of particular importance to USFS. Other information gathered includes the type of area, the priority within the region, acreage, location, price per acre, and total cost. Points are assigned based on whether the project meets needs specified in a forest plan, as well as the OMB criteria. That system was used as a guide to select $100 million of projects from the $500 million of projects that were proposed in FY 1992. USFS fish, wildlife, and recreation staff then decided whether the resulting priorities reflected the national USFS goals.

Interagency coordination generally occurs in response to specific local concerns, such as management of grizzly bears in Yellowstone with USFWS and NPS and a joint effort with BLM to inventory and monitor spotted owls in Oregon. USFS also participates in and uses data from The Nature Conservancy's State Heritage Program.

USFS aggressively pursues increasing the area of land under its jurisdiction. In 1968, it administered 186,893,133 acres (U.S. Public Land Law Review Commission, 1970). By 1991, that figure had increased to 191,324,090 acres (Figure 3-2). In FY 1991, USFS bought 67,321 acres with LWCF money and increased its area by 43,027 acres through land exchanges.

## U.S. Fish and Wildlife Service

USFWS carries out regulatory and land-management responsibilities

**68**        *SETTING PRIORITIES FOR LAND CONSERVATION*

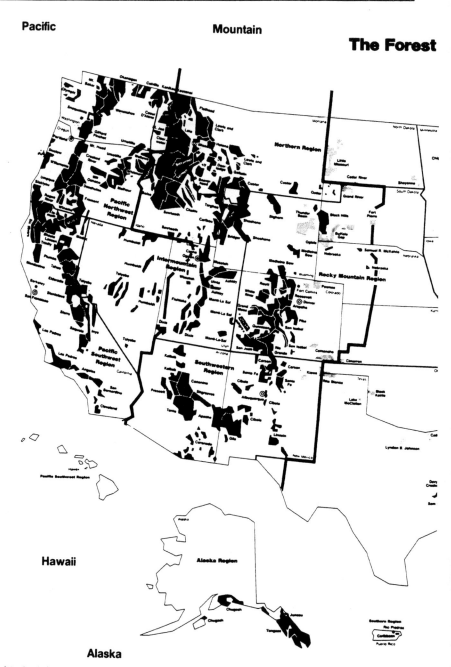

**FIGURE 3-2**    The national forest system.  Source:  USDA, 1991.

# THE LAND ACQUISITION PROCESS 69

## Service

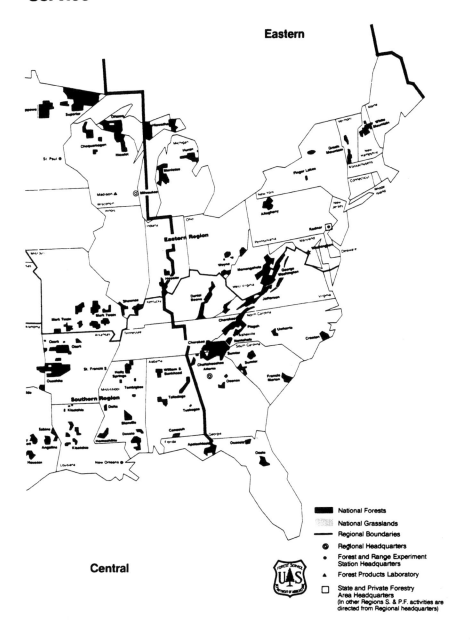

with respect to the nation's wildlife resources. It has primary responsibility for the management of migratory birds, most endangered species, and certain marine mammals. Primary management and conservation authority for other resident wildlife rests with the states, although USFWS closely cooperates with the states in meeting their resident wildlife objectives. For the present study, the national wildlife refuge system is the principal subject of interest.

**Agency Mandate and Acquisition Authority**

The national wildlife refuge system is a set of lands administered by USFWS to conserve the wildlife thereon, prevent extinction, and conserve wildlife ranges, game ranges, wildlife management areas, and waterfowl production areas (Figure 3-3). The system originated with the designation of Pelican Island Refuge by President Theodore Roosevelt in 1903. The biggest impetus for the establishment of a true system of refuges was the enactment of the MBCA in 1929, which authorized USFWS to acquire land or interests in land to conserve the habitats of migratory birds, particularly waterfowl. Much of the system has been acquired primarily for waterfowl conservation purposes, although many units in the system were acquired or reserved from the public domain to conserve endangered species or other designated wildlife.

The national wildlife refuge system is a collection of diverse lands, many managed for unique purposes. The various units were established under many different authorities and for quite different purposes. Many were acquired under the authority of the MBCA "for use as an inviolate sanctuary, or for any other management purpose, for migratory birds." Monies to acquire those refuges come from the Migratory Bird Conservation Fund, comprising the revenues from the sale of migratory bird hunting and conservation stamps.

About 40 units of the national wildlife refuge system were acquired under the authority of the ESA, using monies from the LWCF to conserve one or more threatened or endangered species. Still others were reserved from the public domain pursuant to executive orders or public land orders that described the refuge's purposes. Unlike lands in the park system, only 16 refuges were established through special enabling legislation.

The secretary of the interior is authorized to "permit the use of any area within the System for any purpose, including but not limited to hunting, fishing, public recreation and accommodations, and access whenever he determines that such uses are *compatible* with the major purposes for which such areas were established." That authorization of compatible uses made clear that the national wildlife refuges were to be managed as "dominant use" lands (Bean, 1983).

USFWS makes extensive use of less-than-fee acquisitions compared with other land-management agencies. For example, it acquires a great many easements designed to prevent the drainage of wetlands important for waterfowl breeding in certain regions of the Northern Plains. Those waterfowl production areas are acquired through the Migratory Bird Conservation Fund and typically are not used for recreation or other public uses. Refuge lands acquired through the LWCF, on the other hand, often are administered for recreation and other public uses that are compatible with the refuge's primary wildlife purposes. The acquisition of easements for these types of refuge lands is much less frequent.

From FY 1983 through FY 1991, USFWS received approximately $520 million from the LWCF for land acquisition, an average of nearly $75 million annually. By comparison, from FY 1967 through FY 1982, USFWS received only about $170 million from the LWCF, an average of slightly more than $10 million annually.

## Acquisition Criteria

In response to congressional inquiries about USFWS criteria for determining land-acquisition priorities, USFWS began to develop the Land Acquisition Priority System (LAPS) in 1983. The LAPS manual defines five target areas that are based on agency objectives as specified in 20 primary statutes that include a mandate for land acquisition:

- Endangered species under the authority of the ESA;
- Migratory birds under the authority of the MBCA of 1929, the North American Wetlands Conservation Act (NAWCA) of 1989, and the Small Wetlands Acquisition Program;
- Significant biological diversity under the authority of the Fish and

# 72  SETTING PRIORITIES FOR LAND CONSERVATION

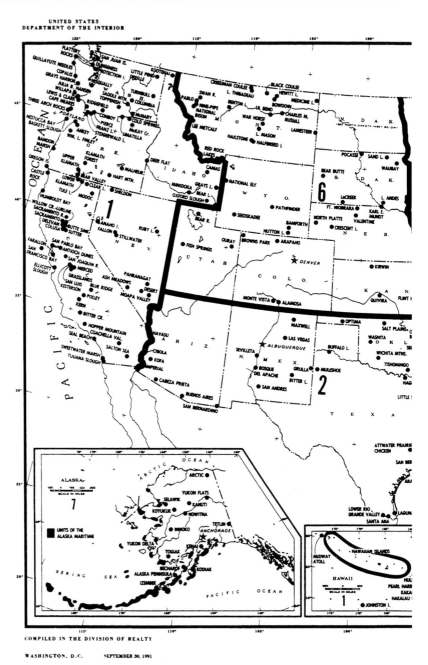

**FIGURE 3-3**  The national wildlife refuge system.  Source:  USFWS, 1991.

# THE LAND ACQUISITION PROCESS 73

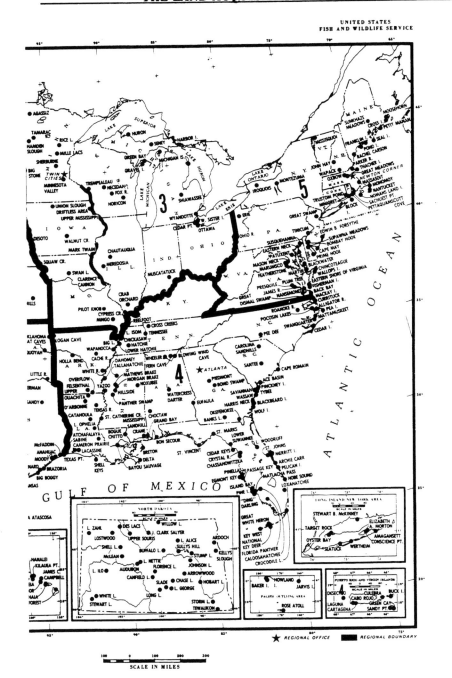

Wildlife Act, the Recreation Use of Conservation Areas Act, the ESA, and congressional recognition of the need to protect biological diversity;
• The nationally significant wetlands target under the authority of the Emergency Wetland Resources Act (EWRA) of 1986, and the NAWCA;
• Fishery resources under the authority of all statutes that require action related to the protection of fishery resources.

Separate criteria were developed for each target and are derived from plans prepared under the different authorities. Criteria for migratory birds are based on the North American Waterfowl Management Plan (NAWMP), for wetlands on the National Wetlands Priority Conservation Plan (NWPCP), and for recovery of threatened species on the published list of endangered and threatened species and species recovery plans. The key concerns expressed in the criteria for endangered species are recovery priorities, species status, and consistency with endangered species priorities (GAO, 1988); for migratory birds, they are habitat loss and population management objectives; for wetlands, they are habitat and threat; and for significant biological diversity they are degree of diversity at various levels, significance of protection, long-term viability, and protection of species of particular management concern. Fishery resources, a new target, emphasizes anadromous and Great Lakes fishery resources represented by indigenous or native species within their original range whose population has been reduced to suboptimal levels as a result of habitat degradation and excessive use.

LAPS has additional criteria common to all projects, including whether a project in one target area contributes to USFWS goals in any of the other target areas or poses threats to the habitat, and if so, permanency and duration of the threats. Other criteria are the percent of the project that would be affected and the potential for public use based on proximity to an urban or tourist area that receives a significant number of visitors.

When a project fits into more than one target area, it is put into the target in which it ranks highest. The regional offices develop an initial list of priorities accompanied by preliminary project proposals for the USFWS director's approval to proceed with the planning process. The preliminary project proposals identify the concerns that would be addressed by acquiring a particular area. Regional acquisition priorities are then compiled in a national data base that ranks projects in accordance with LAPS.

The criteria for merging the target lists, the "budget common factors," are degree of threat, opportunity to acquire land, enhancement of refuge management, acquisition status, development needed to meet objectives, estimated operations and maintenance costs, and change in personnel required. The result is two project lists, one for migratory birds and one for endangered species, that, according to USFWS, best achieve the stated objectives. The USFWS priority lists are then submitted to the department and OMB.

USFWS has condemnation authority and uses it when there is a direct threat to the resource, but USFWS normally does not condemn land except through mutual agreement with landowners to clear a title or when a price cannot be agreed upon. Many projects done under NAWMP are joint ventures with states and local authorities that involve multiple ownership; in multiple ownership projects, USFWS also has to identify the least federal interest that needs to be acquired to achieve project goals (R. Fowler, pers. comm., USFWS, Dec. 3, 1991).

## Bureau of Land Management

### Mandates

BLM was created in 1946 when President Truman combined the programs of the General Land Office and the Grazing Service in a single agency. The General Land Office was responsible for administering the laws that provided for disposal of the federal public domain. These laws had led to the transfer into private or state ownership of two-thirds of the 1.8 billion acres of original public domain. (Some of the disposal laws remained in effect in 1946, although they were used much less then than during the latter part of the nineteenth century.) The Grazing Service was created in 1934 to administer the Taylor Grazing Act, which was to stabilize the western range livestock industry that used the remaining unappropriated and unreserved public domain. BLM lands, mostly in the western states and Alaska, were what were unclaimed after reservations for national forests and national parks, grants to the states, grants to railroad companies for building new lines in the West, and private appropriations for farms, mines, and homesites. BLM was left with responsibility for managing lands that were usually dry, often scattered, and heavily grazed. It was also responsible for managing most of the federal land holdings in Alaska, which included practically all of the state.

Although it did help stabilize grazing on the public domain, the Taylor Act also left the lands and ultimately, BLM, in a difficult position. The act declared that the remaining public domain was to be managed, mainly for grazing, pending its final disposal. At the same time, the act gave the secretary of the interior some authority to limit disposals under the various homestead acts by requiring that applications be accepted only for land that was classed as suitable for the intended use. For many years, BLM was dominated by mining and grazing uses, lacked strong management authority, and risked its land base.

This was corrected to a significant degree in 1976 with passage of the Federal Land Policy and Management Act. FLPMA states that the public lands generally will be retained in federal ownership and managed under the same principles of multiple use and sustained yield that guide USFS in its management of the national forests. But it does not create a system of federal land reservations that parallels the national forests or national parks. This is partly because most of the lands are remnants and pieces of the original public domain. FLPMA did provide for study of possible additions of BLM lands to the national wilderness areas preservation system.

Some interests fully accept the idea that all of these lands will stay in federal ownership. However, after FLPMA was passed, some western states sought to claim ownership of federal lands inside their borders in a "sagebrush rebellion." And a new coalition effort known as the "Wise Use Movement" underscores the recognition of private property rights for those who have timber contracts, mining claims, water rights, grazing permits, and other claims on federal lands (Gottlieb, 1989).

**Acquisition authority**

Before FLPMA, BLM had no general authority to acquire lands by purchase or condemnation (Wheatley, 1970). It did, however, have authority conferred by Section 8 of the Taylor Grazing Act to exchange lands for private lands within the same state or within 50 miles into the nearest adjacent state (Wheatley, 1970). The lands were required to be classified as suitable for exchange before negotiations with another party (Wheatley, 1970).

FLPMA gave BLM its first general land-acquisition authority. It authorized the secretary of the interior to acquire land or interests in land by purchase, exchange, donation, or eminent domain; eminent domain was restricted to securing access to public lands. Congress did not give BLM broad authority to expand the western federal domain.

Chavez (1987) concludes that the authority to acquire access corridors has not solved BLM's problem of affording access to public lands. According to Chavez, the reasons for this are many, but a primary reason is that BLM does not have the funds to acquire easements. For example, landowners sometimes demand prices much higher than a property's appraised value. Some landowners fear that if the public gains access to federal lands, they will lose the revenues earned by selling hunting and fishing rights for their own lands and that the public might harm private lands in crossing to the public lands.

Unlike the Taylor Grazing Act, FLPMA limits exchanges to lands within the same state. This might have been in response to problems caused by some proposed exchanges across state lines—for example, a proposed exchange in the mid-1960s of BLM lands in southwestern Oregon for lands in the Point Reyes National Seashore in California gave rise to controversy within the Oregon congressional delegation and the Department of the Interior (Comptroller General of the United States, 1966; Wheatley, 1970).

FLPMA directs BLM to manage the public lands under its jurisdiction "in a manner which recognizes the Nation's needs for domestic sources of minerals, food, timber, and fiber." But these lands are also to be managed "in a manner that will protect the quality of scientific, scenic, ecological, environmental, air and atmospheric, water resource, and archeological values; that, where appropriate, will preserve and protect certain public lands in their natural condition; that will provide food and habitat for fish and wildlife and domestic animals; and that will provide for outdoor recreation and human occupancy and use." All of this is to be done on the basis of multiple use and sustained yield.

The means for meeting these statutory goals is land-use planning, for which resource-management plans (RMPs) are prepared for each management unit. The content of plans has evolved over the past 15 years. Some of the most recent plans include detailed descriptions of lands that are available for disposal, mainly through exchanges, and lands that are desirable for acquisition. For example, the July 1991 draft of the Judith Valley Phillips RMP for a resource-management area in north-central Montana identifies specific areas for acquisition and disposal (Figure 3-4). The area of BLM land available for exchange in this unit is 166,021 acres, most of which is in scattered parcels of 1 square mile or less. Areas identified for acquisition range from 112,611 acres in one management alternative in the draft plan to 631,719 acres in another. These areas are said to meet acquisition criteria that range from recreation values, riparian-wetland area, and wildlife values to black-footed ferret management (an endangered species) and elk and bighorn sheep habitat in the more expansive management alternatives.

The draft Judith Valley Phillips RMP lists land-acquisition criteria in two categories—general and program specific. Priorities among the criteria are not identified and the range of criteria is broad enough to accommodate most of BLM's program responsibilities. For example, the list includes consolidating of mineral estates and enhancing the opportunity for new or emerging public land uses or values. The accompanying map of acquisition and disposal lands demonstrates that most of the acquisitions would consolidate BLM holdings into more manageable units.

BLM has taken its place with USFS as a major multiple-use management agency. Acquiring lands to provide for public recreation and wildlife-habitat management are signs that BLM actively is trying to meet the evolving needs of land conservation. In pursuing its mission, BLM clearly is trying to consolidate its lands in manageable blocks and using its scattered parcels in trade to accomplish that. That endeavor complements the various conservation objectives that the agency is pursuing with its land-acquisition and exchange program.

FLPMA did not repeal the scattered statutory exchange authorities, but it did expand BLM powers, leading to speculation that the agency should be able to rid itself of difficult-to-manage parcels, consolidate the checkerboard of public land holdings into more efficient units, and accommodate private desires for land transactions with the BLM. Under FLPMA, the lands exchanged must be located in the same state as the interests acquired and should be of equal value. If they are not equal, the values may be equalized with payment, providing payment does not exceed 25% of the total value of the lands or interests transferred out of

# THE LAND ACQUISITION PROCESS

## ACQUISITION CRITERIA

### General Criteria for Acquisition

1. Facilitate access to areas retained for long term public use.
2. Enhance congressionally designated areas, rivers or trails.
3. Facilitate national, state and local BLM priorities or mission statement needs.
4. Stabilize or enhance local economies or values.
5. Meet long term public land management goals as opposed to short term.
6. Be of sufficient size to improve use of adjoining public lands or, if isoloated, large enough to allowed identified potential public land use.
7. Enhance the opportunity for new or emerging public land uses or values.
8. Contribute to a wide spectrum of uses or large number of public land users.
9. Facilitate management practices, uses, scale of operations or degrees of management intensity that are viable under economic program efficiency standards.
10. Secure for the public significant water related land interest. These interests will include lake shore, river front, stream, pond or spring sites.
11. Agricultural lands that would be in the public interest (i.e., management for lure crops).
12. Riparian areas in I and M allotments and important wetland areas.

### Program Specific Acquisition Criteria

*Minerals*
1. Consolidation of mineral estates.
2. Acquisition in response to a federal project need, as in the case of a dam project. Criteria for this type of acquisition would generally include:
   a. Where development of the federal project would preclude the mineral estate owner from exercising development rights, or

**FIGURE 3-4** Judith Phillips resource management plan. Source: BLM, 1991.

**Program Specific Acquisition Criteria** (continued)

    b   Where the exercise of the mineral estate owners right of development would materially interfere with the federal project.

*Livestock Management*
Acquire non-federal holding in I and M allotments, which will enhance manageability and investment opportunity.

*Forestry*
Focus acquisition priority on areas:
1. That exceed 30 cu. ft/acre in growth of commercial timber unless the areas will enhance the harvest of adjacent lands,
2. Contiguous to, or that facilitate access to public forest land,
3. Containing 80 acres or more of commerical timber,
4. Containing enough harvestable volume for a feasible commercial logging unit after physical, biological, or other land use constraints are considered.

*Recreation*
Acquire land with the following signicant values:
1. National values, such as congressionally designated areas, rivers, or trails,
2. State values that enhance recreation trails and waterways or the interstate, state, and multi-county use,
3. Local values for extensive use, such as hunting, fishing, ORV and snowmobile use.

*Wilderness*
Acquire in-holdings within the boundaries of Congressionally designated wilderness areas under BLM administration.

*Cultural Resources*
Any cultural site to be acquired should meet the following evaluation standards: high research value, moderate scarcity, possess some unique values such as association with an important historic person or high aesthetic value, or contribute significantly to interpretive potential of cultural resources already in public ownership.

FIGURE 3-4    Continued

**Program Specific Acquisition Criteria** (continued)

*Wildlife Habitat Management* (continued)
Areas for acquisition will be lands with significant wildlife values as defined below. These areas may be of any size.
1. Threatened and endangered species
   a. Federally listed species
   b. Federal candidate species.
   c. State listed species of special concern.
2. Fisheries.
3. Big game. Important habitat, such as crucial winter areas in I and M allotments with native habitat and associated spring/fall transition areas, kidding/fawning/calving/lambing areas, crucial wallow complexes, mineral licks, and security areas.
4. Upland game birds, migratory birds, and waterfowl. Crucial breeding, nesting, resting, roosting, feeding, and wintering habitat areas or complexes.
5. Raptors. Existing and potential nesting areas for sensitive species or significant nesting complexes for nonsensitive complexes.
6. Nongame. Crucial habitat complexes.

The lands identified in the following table meet the above criteria.

FIGURE 3-4    Continued

**Lands That Meet Figure 3-4 Criteria**
Judith Resource Area
*(Lands Identified Under Alternatives B, C, D, and E)*

| Parcel | Landowner | TWN | RNG | Acres | Reasons for Acquisition |
|---|---|---|---|---|---|
| Black Butte | Private | 17N | 22E | 2,156.97 | Cultural values, hiking, hunting, picnicking, scenic values, sightseeing, access |
| Collar Gulch | Private | 17N | 20E | 280.00 | Camping, fisheries, habitat, fishing, hiking, hunting, picnicking, scenic values, sightseeing |
| DSL Black Butte | State | 17N | 22E | 640.00 | Camping, cultural values, hiking, hunting, access |
| DSL Fergus Breaks | State | 19N | 20E | 640.00 | Excellent mule deer habitat, picnicking, hunting, access |
| DSL Judith-Moccasins | State | 16N | 19E | 460.00 | Camping, hiking, hunting, picnicking, sightseeing |
| DSL Judith Mountains #2 | State | 16N | 19E | 1,816.97 | Camping, hiking, hunting, picnicking, scenic values, sightseeing, access |

FIGURE 3-4 Continued.

**Lands That Meet Figure 3-4 Criteria**

Judith Resource Area
*(Lands Identified Under Alternatives B, C, D, and E)*

| Parcel | Landowner | TWN | RNG | Acres | Reasons for Acquisition |
|---|---|---|---|---|---|
| DSL Judith River North | State | 20N | 17E | 1,460.00 | Riparian, fishing, floating, hunting, access |
| DSL Little Snowy Mountains | State | 12N | 22E | 1,280.00 | Forest management, deer, turkey, sharptail grouse and black bear habitat, hunting, hiking, camping, sightseeing, picnicking, access |
| DSL PN Area | State | 22N | 16E | 2,760.00 | Floating, fishing, hunting, camping, river boat launch area, historical site, riparian, access |
| DSL Snowies Bridge | State | 12N | 20E | 1,280.00 | Forest management, deer, turkey, sharptail grouse and black bear habitat, hunting, hiking, camping, sightseeing, picnicking, access |
| DSL South Moccasins | State | 17N | 17E | 640.00 | Camping, hiking, hunting, picnicking, scenic values, sightseeing, access |

**FIGURE 3-4** Continued.

**Lands That Meet Figure 3-4 Criteria**

Judith Resource Area
*(Lands Identified Under Alternatives B, C, D, and E)*

| Parcel | Landowner | TWN | RNG | Acres | Reasons for Acquisition |
|---|---|---|---|---|---|
| DSL Surenough Creek | State | 14N | 20E | 640.00 | Mule deer and sharptail grouse habitat, hunting |
| Fergus Breaks | Private | 18N | 20E | 2,088.47 | Excellent mule deer habitat, picnicking, hunting, access |
| Judith-Moccasins | Private | 16N | 17E | 2,191.26 | Camping, cultural values, hiking, hunting, picnicking, scenic values, sightseeing |
| Judith Mountains #1 | Private | 16N | 19E | 5,319.04 | Camping, cultural values, hiking, hunting, picnicking, scenic values, sightseeing |
| Judith River North | Private | 20N | 16E | 10,740.00 | Riparian, fishing, floating, hunting, access |
| Judith River South | Private | 19N | 16E | 1,204.16 | Riparian, fishing, floating, hunting, access |

FIGURE 3-4 Continued.

**Lands That Meet Figure 3-4 Criteria**

Judith Resource Area

*(Lands Identified Under Alternatives B, C, D, and E)*

| Parcel | Landowner | TWN | RNG | Acres | Reasons for Acquisition |
|---|---|---|---|---|---|
| Little Snowy Mountains | Private | 12N | 21E | 14,766.50 | Forest management, deer, turkey, sharptail grouse and black bear habitat, hunting, hiking, camping, sightseeing, picnicking, access |

**FIGURE 3-4** Continued.

federal ownership. In the process, provision is made for the appraisal of lands and, ultimately, binding arbitration on the question of valuation. Lands acquired by exchange that are within the boundaries of any unit become a part of that unit (e.g., national forest system, national park system, national wildlife refuge system, and national wild and scenic rivers system). Under later amendments to FLPMA, the respective secretaries were instructed to promulgate "new and comprehensive rules and regulations governing exchanges of land and interests therein."

**Acquisition Criteria**

The LWCF is used to acquire land needed for recreational uses (including cultural), wilderness, natural and scenic areas, fish and wildlife habitat, and critical riparian and wetland areas. Specific planning objectives are specified in BLM's *Recreation 2000* and *Wildlife 2000* initiatives (BLM, 1990; undated). The key objectives in *Wildlife 2000* are to enhance recreation opportunities, acquire critical wildlife habitat, and consolidate scattered tracts of land for more efficient management of resources. Management goals related to threatened and endangered species are to meet BLM's responsibility for recovery of threatened and endangered species on BLM lands and to ensure they are not adversely affected by modification of critical habitat. According to the *Recreation 2000* document, BLM's recreation policies are to provide wide diversity of recreational opportunities and respond to increased recreational demand, provide resource-dependent recreational opportunities, manage and monitor resources essential to recreational experience, use landownership and access adjustments to enhance recreational opportunities by creating more manageable units through consolidation of land holdings, and contribute to local economic vitality through cooperation with tourism entities.

BLM has 150 resource areas and 300-400 management plans (Figure 3-5). Properties identified in land-use planning are screened against national initiatives to give emphasis and priority to noncommodity programs such as recreation and wildlife.

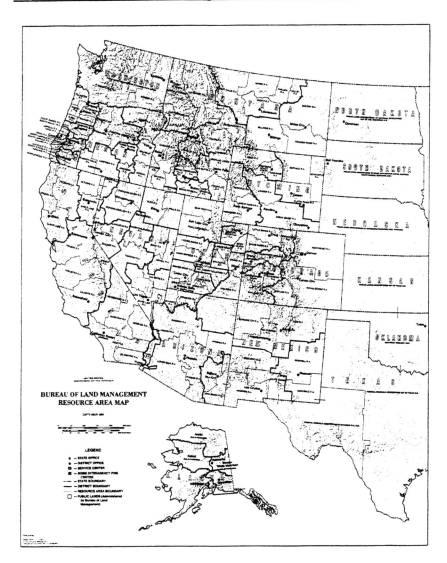

FIGURE 3-5  BLM lands. Source: DOI, 1989.

## OMB Criteria

The need for systematic ranking of land acquisitions among agencies has become apparent for budgeting federal funds, because no cross-agency, national system was available to set acquisition priorities for LWCF appropriations. (President Bush asked in his first budget message that LWCF be funded at an average level of $250 million over 5 years.) To provide a ranking scheme, OMB created a system that incorporates some aspects of the ranking systems of the individual agencies, but also emphasizes the current administration's national priorities. The latter include access to recreational areas close to urban zones and wetlands protection.

To identify acquisitions to be included in the federal budget, each agency's list of priorities is ranked based on uniform OMB criteria, adjusted to reflect department policies, and is submitted annually to the Land Acquisition Working Group.[2] The working group sends its final submission to OMB with written justifications for high-priority acquisitions. OMB criteria are used to establish a single list ranking the requested acquisitions of all the agencies. The budget includes an estimate for the LWCF and a list of acquisitions for each agency for the upcoming fiscal year. The list is then submitted to Congress in the president's budget. Congress reviews and amends the priority list and enacts appropriations for specific acquisitions by agency.

OMB criteria rank potential acquisitions according to a point system (see Chapter 8 and Appendix B). Initially, minimum standards must be met, including

- Availability of the project within the boundaries of or contiguous with an authorized unit;
- Absence of known health or safety hazards;
- Absence of opposition from current owners;
- Limit of 10% of the purchase price for necessary expenditures on the infrastructure (e.g., public facilities, trails, and campsites).

Points are then awarded for roughly a dozen different categories. For example, a parcel slated for development within 2-3 years gets 50

---

[2] The Land Acquisition Working Group consists of the assistant secretary of the interior for fish, wildlife and parks; the assistant secretary of the interior for land and minerals management; and the assistant secretary of agriculture for nature, resources and environment.

points; one slated for development within 4-5 years gets 25 points. This recognizes the danger created by the imminent threat of development which, if allowed to occur, eliminates the potential acquisition altogether.

The criteria can be grouped into five general categories: recreation and access, habitat and wetlands protection, cost minimization, threat of development, and protection of cultural and natural features. The total points a parcel can be awarded for recreation and access is 140; for habitat and wetlands protection, 120; for cost minimization, 70; for imminent threat of development, 50; and for protection of cultural and natural features, 40. Each agency's assistant secretary can add as many as 150 points for that agency's highest priority items, 142.5 for the second priority item, and so on, with the number of points decreasing with the priority assigned to a parcel of land.

The current criteria replace an earlier interagency system to rank priorities among government entities, which was administered by the Heritage Conservation and Recreation Service (HCRS) until 1980. Projects were ranked for minimum, current, and unconstrained budget levels. The criteria were degree of threat, price escalation, and other special considerations. The degree of threat was characterized according to the nature of the threat, the probability of occurrence within 2 years, the severity and permanence of the impact, and the cost of converting land back to original project purposes. Other factors considered were program continuity, congressional directives, and the maximum amount an agency could obligate to a particular area in the given budget year. A line item to purchase inholdings could be used for opportunity buying. After 1980, when HCRS was abolished, the land-management agencies began to function independently.

## THE CONGRESS

Congress has three formal types of direct authority over the acquisition of federal land. First, and by far the most important, Congress annually appropriates funds for land purchases by the administrative agencies. In making appropriations, relevant congressional subcommittees begin with acquisitions budgets proposed for each agency by OMB. Second, Congress authorizes new NPS units; only with this legislative approval may NPS create new units. The other three federal land-management agencies do not need specific authorization to acquire land in new areas. (However, USFS goes to Congress for specific legislation whenever it wants to create a new purchase unit to add to the national forest system.) Third, Congress specifically may forbid agencies to make certain acquisitions, even when the agency is otherwise permitted, or may restrict acquisitions in a given project to transactions with willing sellers or to acquisitions that do not raise the aggregate amount of federal land in a particular county or state beyond a specified level.

Congress also influences land acquisition indirectly. For example, Congress can order agencies to study certain areas (e.g., a potential national park) or certain subjects (e.g., biological diversity). Results of such studies often include recommendations for acquisitions. Congress also controls agency staffing levels; that influences how agencies expand their missions and sometimes, particularly in the case of exchanges, determines how many new projects can be put together each year. Congress can create private entitlements and rights in land, which can necessitate later acquisitions. Finally, through amendments to organic acts or by other legislative directives, Congress can add to or modify agency missions, as it did through the ESA.

For FY 1991, congressional appropriations diverged significantly from administration requests. The administration requested $212 million for 107 sites; of that, Congress appropriated $163 million for 70 sites. Congress then added $110 million for 67 other sites for a total of 137 sites and $273 million. NPS fared considerably better than did other agencies in having its sites approved by Congress: 80% of NPS requests were approved, compared with 64% each for USFS, USFWS, and BLM. Of the sites added by Congress, 30 were NPS acquisitions, 23 were USFWS acquisitions, 11 were USFS acquisitions, and only 3 were BLM acquisitions.

In addition to eliminating some acquisitions and adding others, Congress made changes in requested amounts for individual sites. For example, funding for acquisitions in the Columbia Gorge National Scenic Area was increased to $7.4 million, from an administration request of $4.6 million. A proposal for additional old-growth forest adjoining a wilderness areas on the Mt. Baker-Snoqualmie National Forest received $7 million, although only $2 million was requested. But other administration requests were cut; for example, $10 million requested for the Cumberland Island National Seashore in Georgia was cut by Congress to only $4 million. The frequency with which Congress rejects some administration projects, adds others, and deviates from the OMB rankings provides strong evidence that Congress plays an independent role in the funding process.

## LANDOWNERS

Landowners have considerable influence over the acquisition process in that more than 90% of federal acquisitions are made from willing sellers. In some cases, the sellers are willing only in the sense that they have resigned themselves to federal purchase; in other cases, they are enthusiastic sellers who wish to sell because they support the project or because they want to dispose of the land for personal or economic reasons.[3] Regardless of initial motivation, those who have decided to offer their land for federal purchase are generally eager to have the transaction consummated as rapidly as possible. Federal agencies are aware of this eagerness, as well as the personal and family circumstances that often prompt sales, and try to respond by giving these properties priority for purchase.

Landowners also influence federal acquisition priorities by their land-use decisions. Given the enormous federal purchase backlog, federal agencies generally are content to allow inholdings to remain in private hands, provided land use is not incompatible with the purposes of the federal unit. Owner decisions to cut timber, build structures, or develop mineral deposits frequently will move a property up the federal purchase list. Owners are well aware of this and may announce development plans to encourage immediate federal purchase.

---

[3] Among the enthusiastic sellers are the nonprofit organizations that buy land in anticipation of subsequent federal purchase.

## OTHER INTERESTED PARTIES

Other parties have an interest in the process of acquisition and influence it by exercising political pressure, identifying opportunities, facilitating complex transactions, and identifying important areas for scientific endeavors.

### Nonfederal Governments

Nonfederal governments, including states, counties, municipalities, and special service districts, have three primary interests in federal land acquisition. First, they are concerned with how acquisition changes the tax base. The federal land-management agencies have varied provisions for payments in lieu of taxes (payments that are made to local governments as compensation for tax revenue lost as a result of transferred landownership). Acquisitions sometimes result in net revenue gains to nonfederal governments, sometimes in net losses. These fiscal effects can be critical in determining whether nonfederal governments support or oppose acquisition.

A second consideration is the loss of jobs attendant upon any shift of multiple-use federal lands to biological or conservation reserves. Federal lands can bring in tourists, whose expenditures help a variety of local businesses, a larger federal payroll can aid the local community. But the jobs lost in mining, logging, and milling tend to be high-wage; those who gain and those who lose are not the same people. An periods of painful adjustment may be required if and when offsetting jobs in recreation eventually materialize.

In addition, the dedication of multiple-use land to other uses can damage local communities (which may suffer school, medical service, fire and police closings) by the direct loss of revenues. This is attributable to the loss of the share of federal grazing, mining, and timber harvesting revenues that must be returned to state and local governments under various federal statutes.

The state grants share of LWCF was intended to enable states and local governments to provide recreation opportunities for urban people. This was seen as especially important for the eastern half of the country, which had relatively little readily accessible federal recreation land. For a decade or so after passage of LWCF, federal funds from LWCF were matched with state and local funds to provide and develop recreation opportunities, as well as to protect open space. (Receipt of LWCF funds required that states prepare "Statewide Comprehensive Outdoor Recreation Plans".)

The state grants portion of the LWCF has declined substantially over the past 15 years (Table 3-1). This is the result of budget constraints at all levels of government, as well as growing congressional interest in shifting the balance of funding toward acquisition of federal lands. And without significant federal funds to encourage the states to allocate their limited funds to acquisition and development of recreation lands, the availability of state and local recreation opportunities is not keeping up with population pressures, especially in urban areas.

One result of decreased state funding is a tendency for some states to look to the federal government to provide recreation land and opportunities that would relieve the states of what would otherwise be their responsibility. As described by state officials to the House Energy and Environment Subcommittee of the Committee on Interior and Insular Affairs in a hearing on "The Crisis in State and Local Recreation" priorities for the use of state LWCF funds are for accessible recreational facilities that do not require travel, rehabilitation and refurbishment of existing resources, for enhancement of the state tourism industry—a major employer in many states, and preservation of the rural character of some areas (Travous, 1992). State officials also pointed out that properly placed urban parks are important for connecting the inner city with coastal and wilderness areas.

Lack of funding through the LWCF has also led to other funding mechanisms that complicate the issue. For example, the National Recreational Trails Trust Fund Act, funded by the gas tax paid by recreational vehicle use, was authorized as part of the Intermodal Surface Transportation Efficiency Act of 1991 (P.L. 102-240) and is under the jurisdiction of the Department of Transportation. This has the potential to circumvent the recreational-land-planning process and to result in conflicts with the intent of the LWCF.

## Native American Tribes

The conservation needs of Native Americans have not been considered in the context of the LWCF, because no mechanism has been developed to address the sovereign status of tribes and their role in land conservation. The federal portion of the LWCF is used to acquire public lands; tribal lands are held in trust by the federal government but are not classified as public lands. States have been unwilling to address these needs, because tribal territories are not political subdivisions of state governments. Also, because of the sovereign immunity of tribes, it would be difficult for states to enforce legal agreements upon which grants might be conditional.

In 1979, the Department of the Interior Task Force on Increasing Participation by Indian Tribes in the LWCF prepared a report and proposed legislation amending the LWCF to include tribes. The task force found that tribes have outdoor recreation needs for purposes of meeting community conservation needs and for tourist attraction—just as state and local governments do (DOI, 1979).

Tribes are partners in the federal historic preservation program and are eligible to receive grants for historic preservation purposes (NHPA, 16 USC Section 470). Tribal historic preservation concerns are described in an NPS report on protecting historic properties and cultural traditions on Indian lands and include some concerns similar to those addressed by the LWCF (NPS, 1990). Reservation boundaries encompass approximately 52.5 million acres of land now held in trust by the United States for Indian tribes and individuals (Cohen, 1982). This land is 3% of the land base in the United States and has less than 1% of the nation's cropland, close to 1% of the nation's commercial forest land, and approximately 5% of the nation's rangeland. It also includes significant amounts of U.S. coal and uranium resources and numerous productive oil and gas fields. Native American tribes also occupy important watersheds, particularly in the west and in some of the most pristine areas (e.g., in the areas of Glacier and Yellowstone National parks). The outcome of tribal and state water adjudications could have major effects on what lands can be conserved. Furthermore, with high unemployment rates, tribes face great pressure to develop nonrenewable resources.

The General Allotment Act of 1887 was particularly detrimental to native Americans, because it divided reservation land into small, individual units that could be sold to anyone, disrupting traditional collective landownership practices and use rights and resulting in fragmented land-tenure patterns within many reservations (White and Cronon, 1988;

TABLE 3-2  LWCF Authorized and Appropriated Balances, Fiscal Years 1965-1992

| FY | Authorized[a] | Unappropriated Balance Available | Total Federal Land Acquisition | State Grants Total | Total Appropriations | State % of Total Appropriations |
|---|---|---|---|---|---|---|
| 1965 | $28,398 | $12,398 | $5,563 | $10,437 | $16,000 | 65 |
| 1966 | 109,717 | 0 | 38,429 | 83,686 | 122,115 | 68 |
| 1967 | 95,077 | 0 | 36,207 | 58,800 | 95,007 | 61 |
| 1968 | 112,951 | 0 | 49,093 | 64,038 | 113,131 | 56 |
| 1969 | 253,000 | 88,500 | 116,991 | 47,509 | 164,500 | 28 |
| 1970 | 200,000 | 157,400 | 66,156 | 64,944 | 131,100 | 49 |
| 1971 | 300,000 | 100,000 | 168,226 | 189,174 | 357,400 | 53 |
| 1972 | 300,000 | 38,500 | 102,187 | 259,313 | 361,500 | 71 |
| 1973 | 300,000 | 38,500 | 113,412 | 186,174 | 300,000 | 62 |
| 1974 | 300,000 | 262,277 | 5,480 | 70,743 | 76,223 | 92 |
| 1975 | 300,000 | 254,785 | 121,700 | 185,792 | 307,492 | 60 |
| 1976 | 300,000 | 237,799 | 135,587 | 181,399 | 316,986 | 57 |
| 1977 | 300,000 | 0 | 356,286 | 181,513 | 537,799 | 33 |
| 1978 | 900,000 | 95,000 | 490,880 | 314,120 | 805,000 | 39 |
| 1979 | 900,000 | 257,975 | 360,776 | 376,249 | 737,025 | 51 |
| 1980 | 900,000 | 648,781 | 202,540 | 306,654 | 509,194 | 60 |
| 1981 | 900,000 | 1,260,188 | 108,282 | 180,311 | 288,593 | 62 |
| 1982 | 900,000 | 1,980,261 | 175,546 | 4,381 | 179,927 | 2 |
| 1983 | 900,000 | 2,545,168 | 220,093 | 115,000 | 335,093 | 34 |

TABLE 3-2 Continued.

| FY | Authorized[a] | Unappropriated Balance Available | Total Federal Land Acquisition | State Grants Total | Total Appropriations | State % of Total Appropriations |
|---|---|---|---|---|---|---|
| 1984 | $900,000 | $3,143,278 | $226,890 | $75,000 | $301,890 | 25 |
| 1985 | 900,000 | 3,756,666 | 213,130 | 73,482 | 286,612 | 26 |
| 1986 | 900,000 | 4,488,457 | 120,646 | 47,563 | 168,209 | 28 |
| 1987 | 900,000 | 5,177,831 | 175,656 | 34,970 | 210,626 | 17 |
| 1988 | 900,000 | 5,907,367 | 150,478 | 19,986 | 170,464 | 12 |
| 1989 | 900,000 | 6,601,134 | 186,233 | 20,000 | 206,233 | 10 |
| 1990 | 900,000 | 7,248,775 | 232,599 | 19,760 | 252,359 | 8 |
| 1991 | 900,000 | 7,773,832 | 341,700 | 33,243 | 374,943 | 9 |
| 1992 | 900,000 | 8,352,432 | [b]298,192 | 23,208 | 321,400 | 7 |

[a] An average of 85% of all "deposits" are from OCS revenues.
[b] Estimate.

NPS, 1990). This fractionated land-tenure pattern interferes with the ability of tribes to manage their resources for conservation and survival as Native American communities, because the tribes lack many attributes of jurisdiction over lands inside reservation borders. The American Indian Policy Review Commission found the problem with landownership on Indian reservations to be one of the biggest obstacles to future tribal economic and community development.

Many tribes are seeking to reacquire lands inside reservation boundaries that were lost as a result of the General Allotment Act or through various types of land transactions. For example, on the White Earth Indian Reservation near the headwaters of the Mississippi River, tribal members make up 40% of the population, but only 6% of the acreage is tribally controlled. A community survey was conducted to assess land-use needs and priorities for a land-recovery program. Priorities identified in the survey include land for housing, access to and conservation of wild-rice harvesting and hunting areas, and recreation as well as burial sites and cultural areas. Although White Earth contains several lakefront resorts, few recreational opportunities are available to the Indian inhabitants. The resorts and much of the lakefront properties are owned by nontribal people, many of whom have been identified as willing sellers because of the depressed real estate market and land title problems resulting from transactions earlier in this century.

Sites and areas of tribal cultural, religious, and economic significance are not confined to reservation boundaries, and tribes have an interest in identification, protection, interpretation, and management of these sites as well as access to them. For native peoples, entire landscapes often have historic, cultural, and religious significance (White and Cronon, 1988), including "whole classes of natural elements such as plants, animals, fish, birds, rocks, and mountains" that are incorporated into tribal tradition and help form the "matrix of spiritual, ceremonial, political, social, and economic life" (NPS, 1990). Continued relationships with certain lands and natural resources are key to preservation of cultural heritage as a part of contemporary life and to fighting social problems, such as alcohol and drug abuse, that afflict some Indian communities (NPS, 1990).

## Land-Protection Constituencies

Federal acquisition of land for conservation is advocated by a variety of national, regional, and local constituencies. Some of those constituencies are permanent national-level environmental advocates that seek to protect biological diversity, to expand recreational opportunities, or to accomplish both. Some advocates focus on a particular type of land (e.g., American Rivers Conservation Association) or on land suitable for a particular type of recreation (e.g., Ducks Unlimited). Land-protection constituencies often arrange for scientific testimony in support of particular projects. In some cases, they arrange formal surveys that become the basis for recommendations for setting protection priorities. *The Conservation Alternative*, which is endorsed by 20 national environmental groups, is released once a year. The 1991 book provides an explicit (but unranked) list of recommended federal and state purchases for FY 1992, with a total estimated cost of $1.164 billion.

Local groups sometimes promote particular acquisitions; for example, the Friends of Santa Monica Mountains has been a continuing advocate for completion of acquisition in the Santa Monica Mountains National Recreation Area. Adjoining landowners are sometimes advocates acquisition, either because they share environmental protection objectives, or because they enjoy financial advantages from their proximity to protected lands in the form of leases, appreciating property values or commercial endeavors. For example, tourism-oriented businesses can benefit from creation of new federal recreation areas or even from the publicity attending designation of new wilderness areas or wildlife reserves.

## Land-Acquisition Opponents

Opponents to land acquisition often are advocates of private-property rights and are often landowners concerned about land-use restrictions that might result from land-acquisition and protection strategies (Land Rights Letter, 1991). Other concerns include effects on the local tax base and loss of jobs and revenue from lands used for timber, mining, and grazing. Some opponents also object to federal land acquisition as a matter of principle and advocate the use of conservation incentives for private-property owners or nonprofit organizations.

# THE LAND ACQUISITION PROCESS

One of the top goals of the "wise-use movement," a coalition that includes inholders, is opposition to all use of eminent domain to acquire inholdings. Other opponents fear the loss of access to federal lands for production of commodities; deterioration of rural cultures; federal inattention to maintenance, improvement, and development of existing public lands before acquiring additional land; regulatory land-use restrictions that might result from nearby public acquisition; and potential benefits to certain interest groups and nonprofit intermediaries. Many opponents find intellectual support in the writings of the new resource economists and political support in groups such as the National Inholders Association.

## Acquisition Intermediaries

During the past 2 decades, several national nonprofit organizations have assumed a prominent role as intermediaries in U.S. land protection. These entities sometimes serve as land-protection advocates; sometimes, as owners of properties acquired by purchase or donation, they act as enthusiastic sellers. But they also have a role as facilitators of potential transactions.

The Land Trust Alliance is a network of approximately 900 state and local land trusts that has a major role in project identification and provides an interim source of financing and public support. The Alliance demonstrates an awareness of local and regional concerns and offers understanding of resources and of potential economic effects on the communities in which the individual trusts operate.

Each nonprofit organization has its own criteria for selecting potential land acquisitions. For the American Land Conservancy, the Conservation Fund, and the Trust for Public Land, those criteria largely reflect agency criteria (H. Burgess, ALC, pers. comm., Feb. 13, 1991; P. Noonan, CF, pers. comm., Jan., 1991), because those organizations normally resell properties as rapidly as possible to the government. The criteria of The Nature Conservancy (TNC) are especially well developed and distinctive (see Appendix D).

## Scientific Community

Scientists have been involved in conservation efforts for many years. Recently, they have become involved increasingly with systematically surveying species or ecosystems and determining whether existing protected areas are adequate to ensure their survival. This technique was pioneered in the 1970s by The Nature Conservancy, which undertook "natural heritage inventories" in several states (Hoose, 1981). Today, inventories are available for most states and are being provided in more detail for counties and other substate areas. Detailed information is available on the resources to be found on federal land, after almost 2 decades of planning efforts. Scientists are combining this information with the heritage inventories to prepare sophisticated gap analyses that pinpoint species and systems underrepresented in the federal system. Those gaps become obvious candidates for acquisition (see Chapter 5).

Gap analyses are performed by agency scientists, as well as by acquisition advocates. The resulting lists are the basis for local advocacy of acquisitions as well as for comprehensive proposals such as *The Conservation Alternative*.

## RATIONAL ANALYSIS AND POLITICS IN THE ACQUISITION PROCESS

The differences between agency lists and the projects eventually funded by Congress have led some observers to charge that political considerations override criteria that might be at least described as systematic, and at best as embodying a certain amount of objectivity, rationality, and science.

Stone (1988) observes that "inspired by a vague sense that reason is clean and politics is dirty, Americans yearn to replace politics with rational decisionmaking." Yet, "the enterprise of extricating policy from politics assumes that analysis and politics can be, and are in some essential way, separate and distinctive activities." In fact, politics and analysis overlap in many ways.

Politics is the expression through a variety of governmental processes of group and individual interests. Those interests, in turn, are a complicated mixture of perceptions, ideology, economic self-interest, and altruism. Among the interests expressed in the debate over federal land acquisition are ideological judgments regarding the amount of land that should be controlled by government, local and national preferences for the protection of particular resources or ecosystems, relative weights given to resource conservation compared with recreation, preferences for one sort of recreation relative to another (e.g., snowmobiling versus nature photography), interests of owners of adjoining lands (who might benefit or suffer if adjoining land is federally acquired), and fiscal and economic development interests of local communities. Interests expressed in the political process bring to bear vital information; for example, testimony from landowners can make clear the human costs of a given acquisition, and pressure from urban constituencies can remind policy-makers that there is a demand for urban recreation. Sometimes, political activity even becomes the conduit for scientific information, as has occurred when scientists and acquisition intermediaries inventory natural areas and lobby for their protection.

It should be noted that judgments and values play a role even among scientists and agency professionals (see Hays, 1987). For example, one scientist may value wetlands more than prairie ecosystems, while another may put relatively more weight on creating a natural area system resilient to climate change. A USFS land-acquisition committee, weighing the claims of recreation and habitat protection, might make very different judgments from a USFWS group, even when their scientific competence is indistinguishable.

Foresta (1987) compared the record of national park acquisition during the 1970s in Canada, where parks were selected on the basis of a long-term plan prepared by the Canadian administrative agency—Parks Canada—with the U.S. system, where Congress had the dominant role. Foresta found that 30 units were added to the U.S. park system, but only 4 were added in Canada. All of the Canadian parks were in regions that had been wholly unrepresented in the existing park system. In the U.S., only half were in unrepresented regions.

The politically dominated U.S. park selection system has advantages. Foresta points out that the system is "likely to be more dynamic and therefore capable of protecting many more areas." The U.S. process was also "likely to encourage a greater inventiveness because it allows a wider range of interests into the decision-making process" thereby encouraging a park system that will "serve a wider range of public needs—even though meeting those needs might result in units which take radically new forms or which do not compare aesthetically with the original national parks" (Foresta, 1987).

Congress is clearly the focus of political input into acquisition decisions. But considerations other than those spelled out in agency missions and formal lists of criteria also affect other stages of the process. For example, managers of individual units are well aware of local support for and opposition to particular acquisitions, whether by individuals or by local governments. Agency and staff are not ignorant of congressional preferences. Land-acquisition intermediaries often have continuing relationships with particular agencies or purchase units, so that there is ample opportunity for communication about acquisition priorities.

# 4

# Assessing the Social Effects of Federal Land Acquisition

The successful acquisition of federal lands, whether in the Santa Monica Mountains of California or in the Green Mountains of Vermont, is more than a matter of protecting rare taxa or whole ecosystems, important as these are. Consideration must be given to longstanding ownership interests, social realities, and cultural continuity. Failure to address such interests invites backlash and "sagebrush rebellions" by increasingly vocal and organizationally sophisticated subsets of the American public. An assessment tool for addressing such interests does exist: It is mandated by the National Environmental Policy Act (NEPA) (P.L. 91-190, 42 U.S.C. 4321-4347, as amended P.L.94-83) and calls for environmental and social impact assessment (SIA) when significant federal actions occur.[1]

Between 4 and 5 million acres of private land have been acquired by the federal government through the Land and Water Conservation Fund (LWCF) over the past quarter century. One of the largest annual expenditures occurred in 1985, indicating that even in administrations opposed to federal acquisitions, the American appetite for additional public land continues unabated. As federal landholdings increase, the number of inholders—individuals, groups, corporations, and units of government with property interests inside of federal landholdings—multiplies. The

---

[1]However, see Mandelker, 1984, regarding the ambiguous duty to prepare social impact assessments under NEPA.

growth in the number of inholders is exacerbated by the rapid population growth in counties adjacent to federally designated wilderness areas (Rudzitis and Johansen, 1989) and other federal holdings.

This chapter pursues several objectives. After an overview of the inholder phenomenon, SIA is defined and its procedures outlined. The benefits of such assessments are listed. To illustrate the procedure and its benefits, an SIA case study involving federal land purchase and human relocation is summarized. Finally, the adaptive management potential of SIA for national parks, forests, and other protected areas is discussed. The chapter ends with a look to the direction of public land acquisition and management in the United States and suggests that more social accounting is required if this new direction is to have broad public support.

## INHOLDERS AND FEDERAL LAND ACQUISITION

Inholder concerns are an important part of American federal land policy. The national media frequently report on property owners angered over diminished property rights in and around federal landholdings, and several accounts present the inholder perspective in detail (Arnold, 1982; Williams, 1982).[2] Membership in the National Inholders Association and kindred organizations is on the rise. Scholars have studied the effects of federal land policy on local communities in Washington (O'Leary, 1976), the U.S. Virgin Islands (Olwig, 1980), New Mexico (Knowlton, 1986), in West Virginia (Greer, 1984), in Montana (Blahna, 1986), in Virginia (Perdue and Martin-Perdue, 1979-80, 1991),

---

[2]The Shenandoah National Park offers an example of continuing conflict between the federal government and local landowners in the face of federal acquisitions. The park comprises 196,000 acres in eight Virginia counties. But NPS claims authority to nearly three times this acreage. Were the federal government to prevail, inholder buy-outs and concurrent social effects would follow. For example, 15% of rural Madison County is within the current park boundaries; the county would cede 44% of its land base to NPS at a time when the county's population is expanding. County officials estimate that Madison County would lose nearly $400,000 in land taxes and predict that residents who were relocated when the park was established would need to move again.

and elsewhere. Various investigations point to the need for accommodating the mutual interests of inholders and the public at large in the United States (GAO, 1981; Crespi, 1984; Howell, 1984) and internationally (Rao and Geisler, 1990; West and Brechin, 1991). U.S. history is replete with cases where federal land policy might have harmonized with the needs of local communities had social effects been accounted for. Examples include U.S. Forest Service (USFS) policy toward Hispanic communities in northern New Mexico, National Park Service (NPS) treatment of African-American communities in the Sea Islands of Southeast, the recent efforts of the U.S. Fish and Wildlife Service (USFWS) to purchase wetlands from farmers in New York, the relocation program of the Bureau of Land Management (BLM) under the Hopi-Navajo Land Resettlement Act, and the U.S. Man and the Biosphere Program's attempt to establish the Voyageurs Biosphere Reserve in northern Minnesota.

Inholders and related social issues on federal lands first entered the pages of American conservation history with the creation of Yosemite Park in 1864. Inholder claims to private property rights divided Congress: The House of Representatives supported such rights, but the Senate ardently sought a park without inholders—a view that prevailed after Supreme Court intervention. Despite this early defeat for inholder interests, by 1890 some 65,000 acres of patented lands and 300 mining claims were reported by the U.S. Army captain acting as the superintendent of Yosemite. Thereafter, withdrawal petitions by inholding claimants in Yosemite and elsewhere became "almost a perennial issue on Capitol Hill" (Runte, 1990).

Policy toward inholdings took a new turn in the 1930s when the Taylor Grazing Act all but terminated federal land disposition in the 48 states. Remaining unappropriated and unreserved lands were turned over to the newly established Bureau of Land Management in 1946, and land acquisition programs were begun to augment the public lands of USFS, the War Department, and NPS (Barlowe, 1965). Thus, in the wake of the Great Depression, when a record number of Americans had returned to the land to subsist, federal lawmakers opted to guard what remained of the public domain and add to it from the private estate (Castle, 1982). The number of property owners adjacent to or surrounded by federal holdings grew apace, especially in connection with newly established national parks and forests.

Commenting on this condition in 1946, the NPS director stated that the problem caused by inholding (threat of fire, road construction, and other development) was one of the most serious facing his agency (Drury, 1946). Between 1940 and 1960, the federal reacquisition agenda led to major emphasis on inholder buyouts, a priority carried on in the LWCF legislation (Glicksman and Coggins, 1984).[3] By the early 1970s, however, roughly half the land within the 51 national forests in the Eastern United States remained in private hands (Heritage, 1974), and inholder protests against buy-out strategies surfaced in many national parks, monuments, battlefields, seashores, and wild and scenic river corridors.

As federal holdings have expanded from one-quarter to roughly one-third of the nation's land, inholdings have multiplied. According to the National Inholders Association, inholders number nearly 1.4 million and represent a broad spectrum of American society. Not all inholders oppose federal ownership of land, nor are they necessarily opposed to conservation per se. Inholder concerns and issues, however, can have far-reaching social and political effects that could be addressed if SIA routinely accompanied federal land acquisition.

## SOCIAL IMPACT ASSESSMENT

Social impact assessment is the discovery, comparison, and evaluation of the effects of significant actions before they occur. The effects considered vary from one assessment to another, but include change in residency patterns, recreation use, public health (e.g., noise pollution and physical well-being), transportation, economic well being, and de-

---

[3]With passage of LWCF, Congress agreed that inholdings "ought to be acquired for either their recreational value or in order to improve administration." A substantial part of the LWCF was to be used to purchase such inholdings as one of three original objectives of the act. The 1968 amendments to the LWCF Act permitted the secretary of the interior to acquire privately held lands within the boundaries of national parks in exchange for other federal lands under the secretary's jurisdiction on an approximately equal basis. Under the 1977 amendments, the Senate Committee on Energy and Natural Resources "clearly recognized that the intent of Congress is to eventually acquire all inholdings located in the National Park Service" (Glicksman and Coggins, 1984).

mographic trends. Data are gathered using standard survey techniques, such as questionnaires and interviews. SIA is used to investigate alternative solutions to a proposed action, including no action. Like environmental impact assessments (EIAs), SIA typically is performed during the design stage of an action so that its findings can be applied to implementation phases. The works of Wolf (1983) and Finsterbusch (1985) are widely regarded as definitive guides to implementing SIA.

Section 102(A) of NEPA requires that a systematic, interdisciplinary approach to EIA be used to ensure the integrated use of the natural and social sciences in decisionmaking.[4] Social effects extend to the cultural, economic, psychological, biophysical-health, and political realms (Freudenburg and Gramling, 1992). SIA has demonstrated its effectiveness in detecting and analyzing the social costs and benefits of diverse federal program initiatives. The key findings have been summarized by Freudenburg (1986). Though advisory in nature, well-researched SIA findings are generally accorded importance by interested policy makers.

Several federal agencies have developed useful SIA manuals.[5] Whether the purchase of easements to protect waterfowl habitat, a land swap to consolidate a particular unit, or purchase of full title, SIA offers a methodical, consistent way of identifying the probable effects on private owners, local units of government, and other user interests. SIA is being used internationally to reduce conflict between protected areas and the survival needs of rural populations (Rao and Geisler, 1990; Hough, 1991).

There are compelling reasons to use SIA as a social accounting device when expansion or alteration of federal holdings is proposed. SIA can be used to compare various policy alternatives with the status quo and

---

[4]Environmental impact statements normally focus on the physical and natural environmental consequences of a proposed action and alternatives to that action. Social and economic effects are investigated in relation to those consequences.

[5]USFS, the U.S. Army Corps of Engineers, and the Bureau of Reclamation are three agencies with practical SIA manuals. Most agencies exercise broad discretion over whether to initiate an environmental assessment or a more thorough environmental impact statement, or to exclude assessment for a particular land transaction.

yield better overall land-use planning. Other ways in which SIA can serve the public interest are

- As a management tool, to reduce risk and identify potential problems and expensive mitigations in advance of a significant federal action;
- To generate baseline data against which later monitoring and evaluation data can be compared. Such data allow researchers to discriminate between routine background change and change planned as a result of a particular federal initiative (Ellis, 1989);
- To address the difficult question of who benefits from a transaction, which often is not covered by market analyses, cost-benefit studies, and risk assessments;
- To provide an effective avenue for public participation in important decision-making situations.

SIA does add to the transaction costs of land acquisition. Yet acquisition without SIA is likely to have considerable hidden costs (law suits, unnecessary management outlays, inholder ill will, etc.). Another drawback to SIA is the delay it introduces, which can be a stumbling block during land acquisition. However, often delays to conduct SIA will yield better selection and policy in the long run, despite certain missed opportunities.

## SIA IN PRACTICE:
## A BUREAU OF RECLAMATION CASE STUDY

During the Carter administration, the Bureau of Reclamation initiated a draft environmental impact statement (EIS) on alternative ways to control major flooding of the Salt River in Phoenix, Arizona. An SIA was conducted, because several of the proposed solutions to flooding entailed land acquisitions by the Bureau of Reclamation and the relocation of either the Fort McDowell Indian Reservation northeast of Phoenix or non-Indian area residents within the federal acquisition area. Environmental groups registered concern over flood-control alternatives that would inundate bald eagle habitats and a 4-by-10-mile "garden of Eden" section of the Sonoran Desert. A final EIS was ordered by Secretary of Interior James Watt in 1981.

The resulting SIA was a model effort. Qualified consultants were hired for their technical and scientific abilities and skilled in public participation techniques (Dames and Moore, 1982). Eight alternative plans were investigated. In each, key policy-relevant elements were featured and compared. Each plan was tested for efficiency (objectives met by least-cost means), completeness (meeting all study objectives), public acceptability, and ease of implementation. Particular attention was given to the probable relocation traumas for Indian and non-Indian communities. As with the environmental effects of each plan, mitigation recommendations were set forth, evaluated, and assigned market values to facilitate comparison and decision making.

The social categories of primary interest were cultural-resource loss or impairment (e.g, burial and archeological sites or current settlements), recreation use patterns, public health and safety (e.g., air quality, noise pollution, emotional and physical well-being), transportation changes, land-use compatibility and quality, farmland gains and losses, and a variety of potential economic and energy variables. Those categories were considered at several scales—regionally, locally, and, where relocation would lead to important lifestyle changes, at the individual household levels. Data were gathered from interviews, questionnaires, and participant observation.

Public acceptance of the final flood control and relocation plan was aided by a series of sophisticated public participation processes incorporated into the SIA. A technical agency group consisting of local, state, and federal agency representatives was appointed to assist the Bureau in data collection and analysis. Numerous presentations and briefings were made to special-interest groups, community groups, and area organizations. Public participation was encouraged from the beginning and at key decision points through public workshops and community meetings, educational forums, monthly newsletters and periodic brochures, and other feedback mechanisms. Arizona's Governor Babbitt organized a Citizens' Advisory Committee that met monthly to represent the interests of environment and business groups, the media, Indian tribes, and citizens at large. The committee's final recommendation to the governor conformed closely with the flood-control policy chosen by the Bureau of Reclamation.

Despite additional costs, the public interest was well served by this SIA. The SIA was tailored to the federal acquisition and relocation at hand. Furthermore, the possibility was left open to repeat certain facets of the SIA later in the life of the project, an acknowledgement that social effects change with time and require ongoing evaluation.

## ENVIRONMENTAL MANAGEMENT AND SIA

Lands acquired for preservation and protection are dynamic systems subject to surprise, accident, natural disaster, and value shifts among managers. Consider our oldest national park, Yellowstone. A century ago, an unlimited number of tourists were welcome, hunting was permitted, and predators, such as wolves, were viewed by many as a scourge rather than an integral part of the ecosystem. Even two decades ago, the idea of a Greater Yellowstone Ecosystem, supported with sophisticated satellite imagery and computer models, was unimagined. The thought of Yellowstone geysers being diverted for private thermal power was unthinkable. Status as a national park, forest, seashore, or grassland is clearly not synonymous with unchanging social, technological, and natural conditions.

This reality has long been noted by Holling (1978, 1986, 1992) and constitutes the basis for adaptive environmental management, a process-oriented, "whole project" approach to impact assessment. Social and environmental impact assessments have been modified in the past decade to adopt this contingency-based, longitudinal approach to their subjects. The longer the life of a project or action, the more guarded initial SIA predictions must be and the more compelling periodic replication and restudy becomes.

Social scientists have been receptive to the challenge of adaptive management in SIA. Llewellyn (1974) and Soderstrom (1981) stress the importance of going beyond the preproject emphasis of most SIA research. Wolf (1983) explicitly calls for project monitoring and sustained analysis of effects. Finsterbusch (1985), in seeking greater SIA sensitivity to cumulative effects, similarly extends social accounting beyond the planning stage. Taylor and Bryan (1990) propose an issues-oriented approach to SIA to provide ongoing assessment processes over a project's multiple phases. Freudenburg (1986) calls for SIA procedures that confront the profound turbulence and disorder in society. Elsewhere, Freudenburg and Gramling (1992) articulate the importance of more longitudinal as well as comprehensive SIA strategies.

This same longitudinal perspective is changing the way social scientists perform SIA with reference to conservation and protected area development. Hough (1991) describes the value of long-term SIA monitoring in revealing delayed or unanticipated effects of parks and reserves, which in turn improves planning and protection on-site and elsewhere. Murphree (1991) advocates adaptive management in park creation and management because of the rapid evolution of wildlife use and management strategies, making these a moving target of research. Although Murphree refers to wildlife in the parks of Africa, he could as well be referring to the Greater Yellowstone ecosystem, the Everglades, or Great Basin National Park. The last of these, the nation's newest, allows cattle grazing and is a significant departure from the classical American national parks model. Elsewhere, Geisler (1992) makes the case for adaptive management in the SIAs of protected areas.

## SIA AND CONSERVATION

To the list of benefits accruing to the public from SIA focusing on federal land policy should be added one additional advantage: an improved understanding of the social side effects of the new public-private partnership likely to guide land policy in the future. Before 1979, NPS relied exclusively on eminent domain to acquire land (GAO, 1981). Today NPS and other federal land agencies use numerous acquisition strategies, many of which are less-than-fee. Great Basin allows local stockmen to graze cattle and sheep within the park and to maintain prior water rights. At the same time, tourists, eager to see wildlife such as bighorn sheep or to visit the famous Lehman Caves, number 100,000 per year. Thus, Great Basin follows the Yellowstone tradition of remote scenic grandeur while permitting multiple use.

Conservation perspectives are evolving rapidly in the United States and have, in their new formulation, certain commonalities with the working landscape model in Great Britain (Harmon, 1991), Europe (Beede, 1991), and Israel (Rabinovitch-Vin, 1991).[6] In addition to the

---

[6]The hallmark of the working landscape model is an aesthetic quality resulting from human interaction with the land. The protected area designation is superimposed on an established system of tenure; inholders are an integral part

Great Basin example, biosphere reserves registered with UNESCO's Man and the Biosphere Program are proliferating throughout North America and permit integrated public and private ownership and use activities in their buffer zones. Greenline parks have been recommended by a presidential commission (PCAO, 1988) and rely heavily on easements rather than full purchase of their land base. Current examples include the Columbia Gorge National Scenic Area, the Adirondack Park and the Pinelands Reserve, and the Northern Forest extending from New York to Maine. The Tug Hill Commission of New York is committed to a working landscape model of land management (Tug Hill, 1990), and The Nature Conservancy is partaking in an extensive working landscape experiment in the Hill Country of central Texas (Stevens, 1992). The longstanding multiple-use tradition of USFS and BLM mirrors some aspects of the integrated conservation model of Great Britain as does, from a smaller scale, grass-roots perspective, the adaptable conservation model of the land-trust movement.

---

of the landscape, and although subjected to land-use regulations, are rarely relocated. The majority of land within such parks is privately owned (Harmon, 1991).

# 5

# The Land Acquisition Process and Biological Preserves: A Role for Natural Sciences

This chapter examines several ecological issues that pose challenges to the acquisition of conservation lands and reviews the current state of knowledge regarding those issues. The dynamic nature of landscapes and ecological systems is reviewed at multiple scales, as is the importance of maintaining species and functional ecosystems in lands designated for conservation and the role of spatial configuration in reserve design. The effectiveness of the current set of criteria are evaluated for their effectiveness in addressing the biological components of the agencies' explicit objectives and the issues noted above. Finally, the committee's findings are synthesized and some modifications to the acquisition process are suggested that could enhance the ability to meet the ecological component of acquisition goals.

## FUNDAMENTAL ECOLOGICAL CHALLENGES

### Geomorphic Processes

Geomorphic processes are the physical and chemical processes that determine the distribution of energy within a changing landscape. The distribution of moisture, nutrients, temperature, sediments, and other resources on the landscape affect and are integral to the biological resources, productivity, and diversity of ecosystems. Conservation of biological resources thus entails the conservation of the physical and chemical system—landscape—upon which those resources depend.

The role of water in distributing energy and nutrients in a watershed is central to the viability of ecosystems. Effective conservation of biological resources is dependent upon an understanding of a region's watershed hydrology, water quality, and landscape. The entire landscape may be seen as continously changing in response to changes in the energy regime. The conservation of the landscape as mandated in agency policies thus requires the maintenance not of the present landscape but of the processes that produce and evolve with the landscape over time.

## Hierarchical Levels of Diversity

Biological criteria for land acquisition have evolved with increased scientific understanding, from saving isolated areas for scientific observation to protecting biological diversity and the functioning of ecosystems that involves landscape level considerations. Contemporary thinking in biology also indicates that biological diversity is an important component of sustainable ecosystems. Reasons for protecting biological diversity have been categorized as utilitarian, aesthetic, and moral.

Biological diversity can be conceived as existing in a hierarchy of increasing information or complexity: genetic diversity, species diversity, community diversity, ecosystem diversity, and landscape diversity, culminating in diversity on a global scale. This section discusses strategies for maintaining diversity, while acknowledging that the ecological consequences of losing diversity at and above the species level are not fully understood.

### Genetic Diversity

Populations are genetic pools that differ in the amount of genetic diversity they contain. Conservation biologists believe it is desirable to maintain as much genetic diversity as possible within a species, because species with high genetic diversity generally are less susceptible to extinction. Genetic diversity tends to be higher in species composed of different populations and subpopulations. Therefore, to maintain genetic diversity, a management objective should be to maintain adequate habitats for populations over a wide geographical area. This is particularly important for adapting to rapid environmental change.

**Species Diversity**

Technically, species richness refers to the number of species present; species diversity is a measure of the number of species and their relative abundance. With habitat fragmentation and disturbance, species become increasingly rare. Data must be acquired on minimum viable population sizes and sizes of habitats required to sustain them. For example, Sampson (1980) estimates that populations of the greater prairie chicken can be sustained only on grasslands of 300 hectares or larger and that they also must be within 20 km of other undisturbed grasslands.

**Community Diversity**

This concept generally refers to the number of species inhabiting an area and encompasses all trophic and nontrophic interactions. Many of these interactions are highly species-specific. For example, many species of herbivorous insects depend on specific plant species. Whitcomb (1987) reports that assemblages of leafhoppers in grasslands were dependent on the patch size and structure of host stands, and the rarity of these insects was attributable to the rarity of their host plants. Populations such as these in small, isolated patches are particularly sensitive to disturbance by physical (e.g., fire) and biological forces (e.g., parasites, predators).

**Global-Scale Diversity**

Conservation of diversity must also be considered at a global scale for species with migratory habits (e.g., marine mammals and birds). Also, biogeochemical cycles can be influenced strongly by global phenomena, such as decreasing ozone.

Decisions to acquire land for conservation should be based upon a recognition of the hierarchical levels of diversity, despite the limited data at most levels. Furthermore, acquisition decisions should be based upon clearly stated objectives on what is to be conserved, how, and why. To this end, it is imperative that agencies maintain accurate data bases and maps of what has been acquired. This effort will require more cooperation than in the past among federal, state, and private agencies.

## Conservation-Scale Dependency

Ecological systems and processes can be considered at many different spatial and temporal scales (Allen and Starr, 1988; Delcourt et al., 1983; O'Neill et al., 1986; Turner et al., 1989a). Forested regions can be used to illustrate this concept. Major processes at the global scale, including fluxes of energy, carbon, water, and human activities, such as deforestation and fossil-fuel combustion, influence ecological processes. At subcontinental or regional scales, processes of interest include evolution, extinction, and migration of species and dynamics of disturbance regimes. Land-management and conservation activities are of particular importance at this scale. At the ecosystem level, ecological processes, such as nutrient cycling, productivity, water use, succession, and competition are manifest. Land-management and conservation activities again have an important feedback on ecosystem dynamics. Therefore, the relative importance of processes or feedbacks changes with scale. Observations and conclusions about ecological dynamics also vary with scale. For example, the measurement of pattern in the landscape is sensitive to the resolution and spatial extent of the data being used (Turner et al., 1989b).

The determination of whether a species can persist within an area also might depend upon the extent of the area, because populations occur in a mosaic of habitats. Some areas serve as sources of individuals, where populations produce a surplus, while other areas serve as sinks, where local extinctions exceed births (Pulliam, 1988). Observing only one area leads to different conclusions regarding population persistence and hence, appropriate conservation strategies. Depending on the abundance and spatial relationships of source and sink areas, organisms from source areas might replenish the population from environments where local extirpation was likely.

The factors controlling species distributions differ with scale. For example, in the Southwest, the mortality of oak seedlings at local scales decreases with increasing precipitation, whereas mortality at regional scales is lowest at the drier latitudes (Neilson and Wullstein, 1983). In an area of the Laurentian Great Lakes in Ontario, regional patterns of fish assemblages appear to be determined by postglacial dispersal and lake thermal regimes, whereas environmental conditions such as lake depth and pH assume greater importance in determining species compositions of individual lakes (Jackson and Harvey, 1989).

Finally, a dynamic landscape in which the proportions of different habitat types change through time might exhibit a stable mosaic (Bormann and Likens, 1991) at one spatial or temporal scale but not at another. Without a temporally stable patch mosaic at any spatial scale, fluctuation and change might predominate within areas of any size (Baker, 1989a).

These examples suggest that the importance of spatial and temporal scale as they influence conservation objectives must be considered in the planning for land acquisition. Within a single locality or preserved area, emphasis may be placed on preventing the local extirpation of a species or maintaining a representative habitat type. A particular parcel of land may also be valued for its aesthetic or recreational value. Therefore, conservation goals at the local level may emphasize the perpetuation of a particular ecosystem type. Manipulative management might be required to preserve populations within a local area, but that might conflict with attempts to ensure the perpetuation of broad-scale processes (Baker, 1989a).

The landscape level a mixture of natural and human-managed patches that vary in size, shape, and arrangement (Burgess and Sharpe, 1981; Forman and Godron, 1986; Urban et al., 1987; Turner, 1989). Conservation goals may focus on maintaining a particular juxtaposition of habitat patches. The size and geographic arrangement of patches across a landscape may influence species success or persistence, and many wildlife species are wide-ranging and make use of several habitats (Forman and Godron, 1986). Management or conservation goals might be to perpetuate natural fluctuations in landscape structure (e.g., a natural fire regime), which implies that certain species may fluctuate as well.

If long-term maintenance of biological diversity is a conservation goal, a management strategy that places regional biogeography and landscape patterns above local concerns may be necessary (Noss, 1983). The acquisition of conservation lands would then require an evaluation not only of the habitat within a protected area but also the landscape context in which each preserve exists (Noss and Harris, 1986). When assessing the effectiveness of the criteria by which conservation lands are acquired, the appropriate objectives at multiple scales must be considered.

## Landscape Dynamics

Land conservation is challenging in part because one goal is to preserve areas that are changing (White and Bratton, 1980). When considered over long periods, the species assemblages observed today are relatively recent (Delcourt and Delcourt, 1991). Many have formed only in the past 10,000 years and reflect individual species' responses to changes in the global environment. Over shorter periods—decades to centuries—the patterns of many landscapes are influenced by natural disturbances (White, 1979; Mooney and Godron, 1983; Pickett and White, 1985; Turner, 1987; Baker, 1989b). Disturbances may create openings within forested landscapes, leading to patches of different ages (Runkle, 1985; Knight, 1987; Baker, 1989a,b). Landscape patterns in old-growth forests of New England, for example, result from frequent natural disturbances, such as windstorms, lightning, pathogens, and fire (Foster, 1988). A variety of authors have suggested that natural areas should be sufficiently large to include all normal stages in community development, and that natural processes of perturbation and recovery should be allowed to occur unchecked (Sullivan and Shaffer, 1975; Pickett and Thompson, 1978).

Even in the absence of natural disturbances, landscapes are not static. In the Southeast, for example, forested land has increased during the past 50-75 years after the wide-scale abandonment of marginal agricultural lands (Odum and Turner, 1990; Turner, 1990). In other areas, especially the Midwest, forest cover has declined (Iverson, 1988; Dunn et al., 1991). In addition, many land-management activities (e.g., forestry practices, regional planning, and natural-resource development) involve decisions that alter landscape patterns, with important implications for conservation (Franklin and Forman, 1987). Thus, planning for conservation lands must always assume that the environment is dynamic.

The notion of equilibrium in ecological systems has inspired a long history and controversy in ecology (Bormann and Likens, 1991). The properties that have been used to evaluate equilibrium fall into two general categories: persistence (i.e., simple nonextinction) and constancy (i.e., no change or minimal fluctuation in numbers, densities, or relative proportions). Persistence can be used to refer to all species, as emphasized in many population-oriented models (DeAngelis and Waterhouse, 1987), the presence of all-stand age classes, or successional stages in a landscape (e.g., Romme, 1982). Constancy can be used to refer to the number of species (MacArthur and Wilson, 1967), the density of individual species (May, 1973), the standing crop of biomass (Sprugel, 1985; Bormann and Likens, 1991), or the relative proportions of successional stages on a landscape (Romme, 1982; Baker, 1989a,b).

Studies have demonstrated increasingly either a lack of equilibrium (Romme, 1982; Baker, 1989a,b) or equilibrium conditions that are observed only at particular scales (O'Neill et al., 1986; Allen and Starr, 1988).

Ecological systems can exhibit a suite of dynamics of which equilibrium is but one. The discipline of ecology has gradually adopted a nonequilibrium paradigm in which the dynamic nature of ecological systems and the importance of spatial and temporal scales as they affect conclusions regarding equilibrium states are recognized explicitly. Classic physical theory deals with stability as the monotonic recovery of a system toward equilibrium following a disturbance (May, 1973). This is homeostatic stability in which the system tends to maintain the same state. The field of ecology might require the more flexible definition of homeorhesis (O'Neill et al., 1986): if perturbed, a system returns to its preperturbation trajectory or rate of change. Homeorhetic stability implies return to normal dynamics rather than return to an artificial undisturbed state. Land-conservation strategies must recognize that ecological systems do not exhibit an undisturbed state that can be maintained indefinitely. Rather, ecological systems exhibit a suite of behaviors over all spatial and temporal scales, and the processes that generate these dynamics should be conserved.

### Fragmented Landscapes

A balance needs to be reached between human needs and uses of the land and environmental sustainability (Ruckelshaus, 1989; Lubchenco et al., 1991). Habitats are becoming increasingly fragmented; for example, hardwood forests in the central United States have declined from occupying more than 90% of the landscape in the early 1800s to 20% today (Whitney and Somerlot, 1985). In Onandaga County, New York, only 8% of the area was forested in 1930, and the forest was dispersed in small islands (Nyland et al., 1986). Highways present practical barriers to migration for many species. When suitable habitat for a species is fragmented, the intervening habitat may impede the movement or dispersal of organisms. Changes in habitat connectivity also can influence the susceptibility of the landscape to disturbances, such as the spread of an invading organism (e.g., coyotes and cowbirds).

Consideration of the effect of land-use change and landscape fragmentation on biological diversity is important in designing landscapes that include natural and seminatural areas (Lubchenco et al., 1991), such as the network of lands for conservation in the United States. The acquisition of conservation lands requires an evaluation not only of the habitat within a protected area but also the landscape context in which each preserve exists (Noss and Harris, 1986). Conservation goals may focus on the maintenance of a particular juxtaposition of habitat patches. Whatever the particular goal, conservation must recognize that the regional context, including habitat fragmentation, is important. The challenge is then to optimize the number, size, and placement of preserves across the landscape.

## Climate

Climate is a central controlling factor in all ecosystems and might change in some regions during the next century. The magnitude, rate, and spatiotemporal characteristics of potential changes in temperature and precipitation regimes are not well understood, but studies of historical and prehistoric records make clear that the Earth's climate has been changing on several time scales ever since life began. It is thus almost certain that climate change will continue to occur, regardless of whether it is accelerated by human activities.. In the face of potential environmental change, effective conservation planning must be based on an understanding of the major biotic and abiotic controls on the ecological components and processes of the area; an expectation of how these controls may change, either through natural events or human actions; and an evaluation of the probable effects of these changes on the ecological processes and biota of a region (Golley, 1984).

*Disturbance regimes.* The frequency, duration, and severity of abiotic and biotic disturbances are likely to be altered by climate change. For example, forest-fire frequencies should increase where the climate becomes warmer and drier (Sandenburgh et al., 1987). Patterns of biotic disturbances might be altered. Because their ranges are often limited by climatic factors, the distributions of pests or pathogens might change with climate.

Short-term climatic fluctuations provide important insights into the response times of species and landscape mosaics to rapid environmental changes in disturbance regimes on the order of decades to centuries. For example, in northwestern Minnesota, changes in the charcoal influx to lake sediments demonstrate how alternating periods of cool-and-moist cycles and warm-and-dry cycles since 1240 A.D. have influenced the periodicity of fire (Clark, 1988).

If ecological disturbance regimes are altered, changes are likely in many landscapes. Habitat types might be eliminated locally from certain areas. Thus, conservation planning must consider whether the size of an existing or proposed reserve in a disturbance-prone environment would be adequate to incorporate an alteration in disturbance frequency or severity. In addition, if an altered disturbance regime could lead to the loss of some habitats, the regional context of the reserve and the potential for the persistence of the desired habitat in other geographic locations should be evaluated.

*Changes in the location of suitable habitat.* A second effect of climate change might be a gradual movement of potentially suitable conditions for different species. Species would be expected to migrate to hospitable environments. However, migration rates are difficult to predict, because the rates of climate change are not predictable. Furthermore, there are now new barriers to migration (e.g., cities, agriculture, and roads) and new modes of migration (e.g., cars, trains, transplants for horticulture, forestry, or agriculture.) Range extension in the future may be less efficient than in the past, because advance disjunct colonies have been extirpated by human disturbances, and propagule sources often have been reduced (Davis, 1989a). The current spatial distribution and abundance of a species will influence its ability to migrate successfully to regions of suitable climate and soils (Peters and Darling, 1985).

A variety of modeling approaches have been used to explore the potential redistribution of habitats in response to changes in climate and atmospheric $CO_2$ (Davis and Botkin 1985; Emanuel et al., 1985a,b; Solomon and Webb, 1985; Solomon, 1986; Pastor and Post, 1988). Simulation results suggest that species abundances are not always in equilibrium with climate; biotic interactions or other environmental factors (e.g., soil heterogeneity) can have strong effects and may obscure or delay observed responses; and lags of as much as 1,000 years may occur in biome shifts. Time lags in species responses to historical climatic changes have been documented empirically. For example, beech has animal-dispersed seeds and tends to move as a front. Beech showed a time lag of 500 to 1,000 years in crossing from the eastern to the western shore of Lake Michigan (Davis, 1989b). In contrast, hemlock, whose wind-dispersed seeds can travel 100 km beyond the main species front, showed no time lags attributable to crossing the Great Lakes (Davis, 1989b). The most common species will be dispersed to new habitats by humans, but time lags will be a problem for unmanaged forests, natural areas, and preserves.

The important point regarding the planning of conservation lands within the United States is that the long-term persistence of some species will be influenced by the availability of migration corridors and new locations of suitable habitat. This strongly suggests that, when possible, the regional-scale connectedness of natural habitats be considered in evaluating potential acquisitions.

## Reserve Configuration and Landscape Linkages

The effectiveness of an area protected for conservation is a function of its size, shape, and connectivity to other sites. However, optimization of reserve design remains a topic of research. Frankel and Soulé (1981) suggest that the rate of species loss increases with decreasing reserve size. Indeed, the positive relationship between species richness and area is well established. One analysis (Newmark, 1986, 1987) showed loss of certain kinds of mammal species from western national parks. Current analyses of data on birds, butterflies, and small mammals from the Minimum Critical Size of Ecosystems project suggest the importance of large areas (Lovejoy et al., 1986). Nonetheless, large and small are relative terms. A different approach is to examine the population of the lowest density species (grizzly bears or grey wolves in Yellowstone, for example) and determine the area necessary to support what could be considered a minimum viable population. This suggests that maximizing reserve size is desirable for maintaining species richness. It must be recognized, however, that even the largest nature reserves, if left alone, will probably suffer major die-offs of species in a few hundred or a few thousand years (Frankel and Soulé, 1981). For example, the huge Kruger National Park in South Africa—about 350 km long and 80 km wide in places—requires significant management to protect many of its species from major population declines and perhaps even extinctions (Aiken, 1988).

Even if a habitat fragment is suitable to support a population of interest, there is no assurance that the population will remain viable if it is isolated from other populations, because genetic variability may be lost through inbreeding. The shape of a reserve and "edge effects" influences the relative amount of edge to interior, which in turn influences biological diversity and susceptibility to disturbance (Burgess and Sharpe, 1981; Ranney et al., 1981; Harris, 1984, Lovejoy et al., 1986). A circular area, for example, has the lowest edge to area ratio, whereas a long, thin rectangle has a much larger edge to area ratio—in fact, it may be all edge.

For decades, the interspersion of habitat types, the creation of edge effects, and the juxtaposition of different kinds of plant communities were believed to enhance wildlife habitat values (Harris and Scheck, 1991). It is now recognized that the creation of distinct edges (e.g., clear-cuts next to old growth) may reduce the biological value of an area by increasing susceptibility of undesirable disturbances (Franklin and Forman, 1987).

The effects of habitat fragmentation and connectivity have been studied extensively through empirical studies and models. Milne et al. (1989) found that wintering white-tailed deer did not use sites containing suitable habitat that were isolated from other suitable sites. In northern Florida, approximately half of the breeding bird species characteristic of hardwood forests do not reproduce in small forest fragments occurring in an agricultural matrix (Harris and Wallace, 1984). In examining the effects of alternative arrangements of corridors of "equal strength" in a simulation model, Lefkovitch and Fahrig (1985) demonstrated that populations in isolated patches experienced earlier local extinction and had lower average population sizes than patches connected by corridors to other patches. In addition, patches that formed part of a square or pentagon had higher population sizes and probability for survival than those that formed part of a line or triangle. These results suggest that corridors or links between habitat patches can be a major factor in the long-term survival of a population; however, such linkages may also enhance the spread of a pest species.

It is generally accepted that the destruction and fragmentation of habitat is the most important cause of species loss. Landscape designs that facilitate movement and dispersal of native biota are preferred by conservationists over designs that do not (Soulé et al., 1988). Moreover, because there is already an established system of protected areas, the function of existing conservation lands would be enhanced by conservation strategies that emphasize movement and dispersal. While corridors may be beneficial to species and be critically important in mitigating the effects of climate change, the role of corridors as an aid to dispersal will vary by species. However, the presence of corridors should not be used to justify the establishment of smaller reserves (Frankel and Soulé, 1981).

There are some examples of land acquisition to link protected areas across the landscape. According to Harris and Scheck (1991), 80% of Florida's largest protected natural areas are too small to contain a single pair of wide-ranging species, such as the Florida panther or black bear, and about 60% of the largest 315 areas are less than 400 hectares. The state of Florida, in conjunction with The Nature Conservancy, is in the process of acquiring land to link existing protected areas. Acquisition of the Pinhook Swamp on the Florida-Georgia border creates a federally owned protected area of 225,000 hectares that extends nearly 100 km (see Figure 5-1). Another project in Florida aims to consolidate several state and federal parks and protected areas along the Wekiva River north of Orlando. This area will span more than 150 km through the Ocala National Forest and along the Oklawaha River (see Figure 5-2). In the Southwest, USFWS, the Texas Parks and Wildlife Department, and several nongovernmental agencies are cooperating to develop the Rio

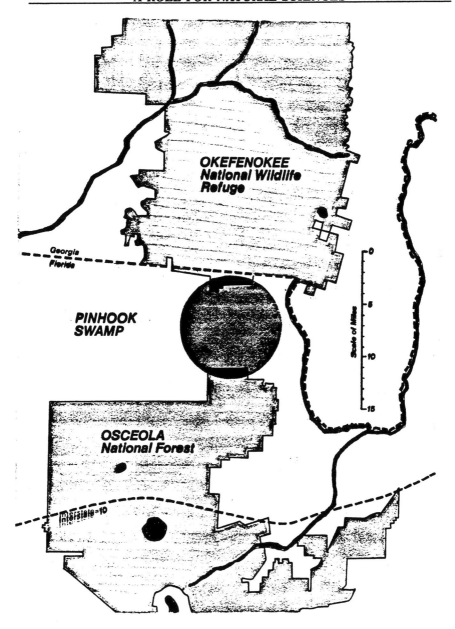

**FIGURE 5-1** Okefenokee National Wildlife Refuge/Osceola National Forest linkage. Circle is linkage area. Source: Harris, 1988.

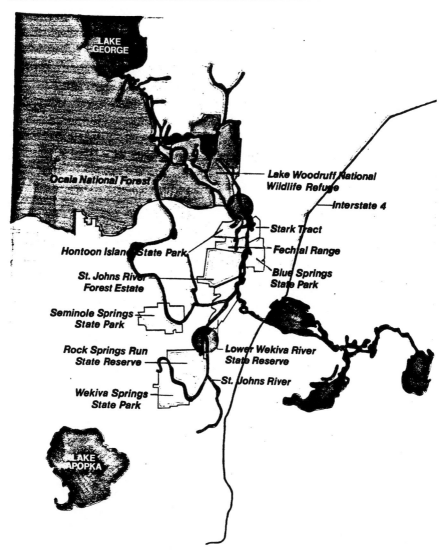

**FIGURE 5-2** Ocala National Forest/Lake Woodruff National Wildlife Refuge/state parks linkage. Circles are linkage areas. Source: Harris, 1988.

Grande Wildlife Corridor (Harris and Scheck, 1991). This effort would link several dozen existing and proposed protected areas with interconnecting corridors that will span 750 km.

Other schemes for corridors have been proposed, including acquiring defunct railroad rights of way to connect habitats.

## OTHER BIOLOGICAL CONSIDERATIONS

Biological criteria fall into three categories. The first involves the choice of the biological feature (objective) to be conserved; this is generally a matter of a particular habitat type or set of biological diversity. The second is reserve design, essentially to ensure the objective can be realized over at least centuries if not millennia and assuming only natural environmental change. The third category focuses on extrinsic and anthropogenic factors, which if ignored, could threaten the best chosen and designed of biological reserves.

Protection of biological diversity is the fundamental purpose of the state heritage programs established in most states by The Nature Conservancy and of the USFWS's Geographic Information Service (GIS) program, although both programs are limited by the available information on biological diversity, which is uneven in depth and recency.

It is possible in a general way to ensure that a representative set of vegetation and other habitat types is under some form of protection. Although extensive debate about a classification of territorial habitat types is possible, such a classification is much easier than defining a set of unique aquatic habitat types. Nonetheless, clear priorities even arise in aquatic habitats, such as the drainages of the Southeast, which are characterized by many endemic species (particularly molluscs). On land, major gaps are evident, such as the absence of tall-grass prairie on federal lands where the primary purpose is habitat protection.

Clearly, an ongoing process of biological inventory and monitoring is needed. This should build upon existing programs, such as biological surveys and heritage programs, the USFS research natural areas program, and the knowledge banks represented by collections held by botanical gardens and natural history museums.

Once a biological feature is selected for protection by land acquisition, attention must be paid to reserve design. This can be a relatively simple question if it is a fairly distinct, cleancut feature, such as a bog. The delimitation of bog vegetation is easily determined. If, however, the objective is to protect a representative sample of biological community that is widespread, the question of size immediately arises.

Wherever possible, habitat chosen for acquisition should not be fragmented. Rather, habitat should be continuous but appropriately should include natural disturbance regimes. When it is not possible to find an area of continuous habitat that is large enough, two solutions are evident: acquire the necessary area and encourage the return of natural vegetation between the fragments, or ensure the core area is surrounded by a matrix of habitat fragments and corridors that provide for species populations larger than the actual protected area can.

The best-chosen and best-designed area for biological diversity conservation is nonetheless vulnerable to outside factors. Total watersheds and their management need to be taken into account in design and management. Otherwise problems may result, such as the toxic agricultural runoff that poisoned waterfowl in the Kesterson Wildlife Refuge. Acid rain emanating from anthropogenic sources far from a reserve can alter lake acidity and even growing patterns or survival of trees.

While the extent, rate, and details of climate change introduced by artificial release of greenhouse gases might be a matter of disagreement, it could have a major negative effect on biological diversity (Peters and Lovejoy, 1992). Few measures can be taken to avert such adverse effects. When possible, the best measure is to conserve altitudinal gradients that will allow species to move upslope in the event of temperature increase. If altitudinal gradients are not available, latitudinal gradients can be considered, but this involves much more extensive expanses of natural habitat. This assumes of course that the only climatic change will involve temperature as opposed to rainfall, snowmelt, directions of currents, etc.

## ENHANCING THE ECOLOGICAL EFFECTIVENESS OF THE ACQUISITION PROCESS

### Gap Analysis

Past land-acquisition strategies have focused on saving critically endangered species and their habitats, rather than protecting ecosystems or landscapes at a broader scale to prevent species from becoming endangered (Noss, 1987; Scott et al., 1988). New conservation lands that consider biological diversity require answers to a variety of questions, some of which have been summarized by Scott et al. (1988) and Davis et al. (1990):

- Are existing preserves located in areas of high species richness?
- Are threatened, endangered, or other species of special interest represented in protected areas?
- What species do and do not occur in protected areas?
- What proportion of threatened, endangered, or sensitive species is protected in existing preserves?
- What is the range of a given species or community type?
- What are the trends through time in diversity at all scales?
- Which ecosystems are adequately protected?
- What is the ownership of species-rich areas?
- Which unprotected areas that are biologically important are at greatest risk?
- How will changes in land use affect the number of species not found in protected areas?
- Where can ecologically sustainable development occur with acceptable impacts on biological diversity?
- Do adequate landscape corridors exist between areas of high species richness to provide for dispersal and interbreeding of populations?

Biological diversity assessments frequently are based on either species or communities. Vegetation types often are used as indirect indicators of the distribution of biological diversity (Diamond, 1986; Backus et al., 1988; Crumpacker et al., 1988; Huntley, 1988). For example, the U.S. Man and the Biosphere Program attempts to designate biosphere reserves within 24 biogeographic provinces in North America (Udvardy, 1984) to ensure that representative areas within each province are conserved.

Methods are needed for evaluating the present status of conservation of biological diversity within a region and identifying unprotected areas that would enhance protection status. Gap analysis is a method for identifying biological diversity preservation needs by analyzing gaps in the present network of protected areas. The approach entails examining the distribution of several key elements of biological diversity relative to areas now under some type of protective ownership. As proposed by Scott et al. (1988), gap analysis requires the following information: (1) the actual vegetation types, recommended at a scale of 1:250,000; (2) the distribution of terrestrial vertebrates, including centers of species richness for different taxa in each vegetation type and biogeographic province, centers of endemism, and the status of protection of individual species; (3) the distribution of terrestrial invertebrates (especially butterflies), including centers of species richness for different taxa in each vegetation type and biogeographic province, centers of endemism, and the status of protection of individual species; (4) the distribution of threatened, endangered and sensitive species; and (5) the distribution of areas managed for the preservation of biological diversity, including public and private nature preserves, with an assessment of the degree of protection offered by present management.

The output of a gap analysis is a map of candidate sites of high conservation value and tables of communities that are not represented or are under-represented in a network of reserves (Davis et al., 1990). Gap analysis to determine which species or habitats occur in protected areas and where additional protection might be most effective is much more effective when the data are stored and analyzed in a GIS (Davis et al., 1990). Scott et al. (1987) demonstrated the power of this approach by comparing the species richness of endangered forest birds in Hawaii with existing reserve boundaries, demonstrating that protected areas in Hawaii generally did not coincide with the locations of the endangered birds (see Figure 5-3). Digital GIS can be a more effective approach than either manual methods or nonspatial automated means of making an assessment of biological diversity (Davis et al., 1990).

Gap analysis can be applied at several scales, from within a state or subregion of a state to a country or continent. In addition to identifying the gaps in the network of protected areas, gap analysis can also be used to monitor the effects on biological diversity of public management practices (e.g., patterns and intensity of resource exploitation) as they alter or fragment habitat (Scott et al., 1988). Gap analysis programs are in place in Oregon, Utah, Idaho, and California. Conducting a comprehensive gap analysis is an ambitious undertaking, but it appears technically feasible to do so at least statewide (Davis et al., 1990), and Scott and Csuti (1992) estimate a nationwide program would require about 6

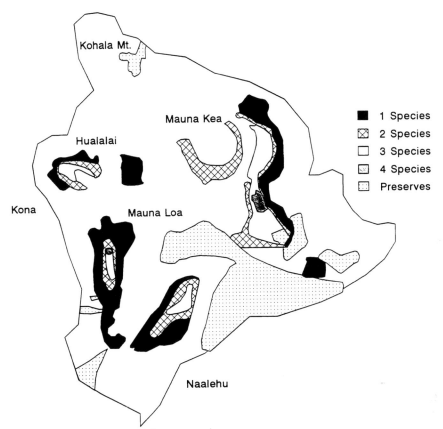

**FIGURE 5-3**   Gap analysis of Hawaii. Source: Scott, 1991.

years to complete. It is not clear whether gap analysis could be applied to aquatic systems.

## Geographic Information Systems

The geographic information system (GIS) is a powerful tool to plan for and acquire lands. The power of GIS lies in the ability to manipulate and analyze spatially distributed data (Figure 5-4). A GIS consists of the computer hardware and software for entering, storing, transforming, measuring, combining, retrieving, and displaying digitized thematic

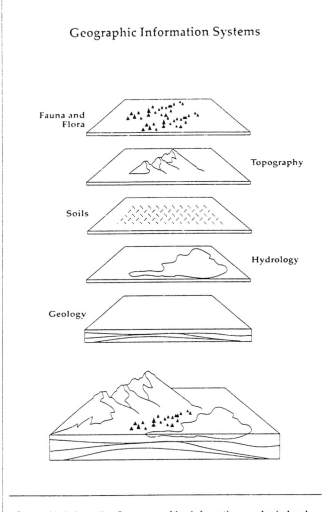

**FIGURE 5-4** Layers of GIS. Source: Reprinted by permission of the publishers of *The Diversity of Life* by Edward O. Wilson, Cambridge, Mass." The Belknap Press of Harvard University Press. Copyright 1992 by Edward O. Wilson.

data that have been registered to a common coordinate system. Because the data can be accessed, transformed, and manipulated interactively, they can serve as a testing ground for studying environmental processes, analyzing the results of trends, or anticipating the possible results of planning decisions (Burrough, 1986). Planners and decision makers can use GIS to explore a range of possible scenarios (e.g., alternative arrangements of conservation areas) and evaluate potential consequences of a course of action before changes are made in the landscape. Importantly, quantitative assessments can be conducted over a broad range of spatial and temporal scales.

GIS is used widely in urban and regional planning, natural resource planning and management, and landscape architecture (Johnston, 1987; Ripple, 1987; Johnson, 1990). Forest managers routinely use the inventory capabilities of GIS at the federal level (Chambers, 1986), state level (Tosta and Davis, 1987), and local level (Wakely, 1987). GIS also is being used to explore the implications of land-management alternatives. Application of GIS to aquatic conservation efforts is more difficult, although it has been used, for example, to establish buffer zones around rivers to determine how land use would change water quality (Johnston et al., 1988; Osborne and Wiley, 1988). GIS systems are easily linked with remote sensing imagery, and linkages with simulation models are being developed rapidly (Coulson et al., 1991).

A recent review of GIS applications in natural resources and ecology (Johnson, 1990) highlights several important operations that are relevant to conservation planning. First, GIS can be used to determine the spatial coincidence of different types of spatially distributed data. Coincidence analyses result in digital maps showing the areas of overlap between two or more data layers (e.g., soils and vegetation). For example, GIS and remote sensing imagery have been linked successfully to predict the occurrence of species populations based on the coincidence of required habitat and environmental factors (Scepan et al., 1987; Stenback et al., 1987; Hodgson et al., 1987) and to identify potentially suitable sites (Palmeirim, 1988; Milne et al., 1989). By selectively weighting habitat characteristics and describing spatial variables such as patch size, shape, and arrangement, the quality and quantity of habitat can be estimated (Johnson, 1990). Second, temporal changes in landscape patterns can be quantified using GIS (Iverson, 1988; Turner, 1990). Although temporal analyses have emphasized the detection of landscape or regional change, the approach could be applied at a variety of scales. GIS could be used to identify areas undergoing the most rapid changes in which conservation needs might be most crucial. Third, proximity, or neighborhood analyses, are used to examine spatial interrelationships. For questions of species persistence, interpatch distances or distances to multiple required habitats could be computed for different conservation strategies. Distances to various features, such as roads, urban areas, or recreation sites, can be computed easily.

GIS is becoming indispensable in land management, and many federal agencies have developed or are developing geographic data bases for their lands. The availability of digital data and the analysis capabilities of GIS could significantly enhance the process by which alternative acquisition strategies for conservation lands are evaluated.

Responses to an inquiry made to the agencies by the committee demonstrate considerable interest and progress in establishing GIS systems on lands managed by each agency. The present status and use of GIS varies substantially among the agencies, but the extent of coordination among the agencies is not clear. Conversion between alternative GIS software programs is becoming increasingly simpler, but the inherent limitations in the scale of the original data incorporated in the data base can limit the ability to combine different data bases.

## Landscape Pattern Analysis

The evaluation of conservation lands within a regional context requires quantitative methods to describe the regional landscape. New methods to analyze and interpret landscape patterns have developed rapidly during the past decade. Those methods include overall indices of landscape pattern, such as diversity, evenness, dominance, contagion, or fractal dimension (Romme, 1982; Krummel et al., 1987; Milne, 1988, 1991; O'Neill et al., 1986, Turner, 1990; Turner and Gardner, 1990), as well as measure of patch characteristics (e.g., number, size, shape, and interpatch distances) or frequency distributions (Gardner et al., 1987, Baker, 1989b).

Landscape pattern analysis typically is based upon land-use and land-cover data, which include categories such as forest, cropland, rangeland, urban, wetland, and water. A commonly used classification scheme

(Anderson et al., 1976) is hierarchically based—analyses can be conducted either with broad classes (such as forest) or more detailed categorizations (e.g., deciduous, coniferous, or mixed forest at next level). Depending on the purpose, more finely divided classes of vegetation or habitat types could be used, as could human uses of the land. Landscape data generally assume some level of homogeneity within each category that is mapped in the landscape. Thus, categories are selected with regard to a particular question or purpose. For example, to study landscape patterns in the East, general categories would be most appropriate, whereas to study vegetation patterns related to prior land use in Great Smoky Mountains National Park, different vegetation classes would be most appropriate.

The selection of the categories used in an analysis places constraints on the interpretations. If digital data are available in a GIS, landscape pattern analysis is straightforward and can provide valuable information.

Quantitative measures of landscape patterns serve several purposes that are relevant to conservation lands. The abundance, size, and shape of habitats within a conservation area can be determined, and changes in those attributes through time can be monitored. In addition, the status of different habitat or vegetation types within a region also can be evaluated and monitored through time.

## CONCLUSION

Ecosystem management has been fragmented because many owners and agencies have responsibilities for different ecosystem components. Preservation of biological diversity also has been thwarted by lack of landscape-level management, insufficient data, and competition among federal land-management agencies (Grumbine, 1990). Although ecosystems cannot be protected using land acquisition alone, acquisition decisions should be made to promote long-term sustainability of the regional resources whenever possible.

No single approach or level of analysis can address all conservation issues. Conservation of biological diversity has biotic as well as abiotic dimensions; protection should consider the interrelationships among the different scales of biological diversity, including gene populations, species, community, ecosystem, biome, and biosphere. The social institutions created to manage ecological resources (including the federal land-management agencies) accomplish their objectives in a variety of ways, ranging from day-to-day management decisions at individual sites to agencywide decisions, such as assigning priorities for acquiring lands for conservation purposes. Each agency has a unique combination of missions, traditions, policies, and explicit priorities that drive its land-acquisition policies. In principle, that diversity facilitates a variety of conservation strategies ranging from total protection of pristine lands to the acquisition of lands for intensive management and manipulation for specific objectives. But a critical question is whether this system gives adequate consideration to conservation issues that transcend the jurisdictional boundaries of the federal land-management agencies.

Many additional important conservation issues are becoming apparent to the scientific community. A few examples are climatic change, declines in populations for entire groups of species, and fragmentation and insularization of biotic communities. Natural ecosystems are spatially and temporally dynamic; furthermore, the importance of a single site to regional biological diversity is variable. The tendency, however, has been to establish geographically fixed wildlife refuges with immovable borders that inhibit species survival.

Rivers are the products of their drainage basins, and the biological integrity of stream and river systems is dependent to a large extent on land uses and management practices in the entire watershed. Such land uses, as well as outside factors, such as demographics, must be considered to identify and protect critical areas. Long-term planning is needed to develop a land-acquisition strategy that considers acquisition projects in the context of the landscape on a scale appropriate to the needs of affected plant and animal species. A recent Naitonal Research Council report on restoration of aquatic ecosystems also addresses landscape considerations and suggests that long-term planning be done by regional planning programs organized by watershed basin (NRC, 1992b).

The ultimate goal of conservation activities should be sustainability of renewable resources, including cultural and biological dimensions. Addressing the problem of sustainability requires an interdisciplinary, complementary approach among natural and social scientists, as well as cooperation among agency programs with diverse and often conflicting mandates. Land-acquisition programs can play a significant role in achieving the goal of sustainability through the acquisition of critical areas.

Achieving sustainability requires a recognition of practices that are not sustainable, an understanding of the capacity of watersheds to support multiple uses, and the identification of tradeoffs between meeting current needs and maintaining a diversity of options for the future. More attention is needed on urban ecosystems, where the majority of the human population resides, and their relationships to agricultural, forest, and natural terrestrial and aquatic ecosystems that make up the cultural landscapes of the biosphere (Huntley et al., 1991). A document that resulted from a workshop, "International Sustainable Biosphere Initiative," notes that:

> Multiple use of ecosystems has long been the norm in human societies. However in the past, usage patterns tended to evolve over a longer period of time and impacts were of lower intensity. Thus decisions were often reversible and often did not limit the options of future generations. Now, impacts are greater in intensity and area, and organizational structures often make it more difficult to achieve consensus on usage patterns. Thus, there is a need for new decision support systems. These systems must bring together information on physical, biological and socio-economic domains.

The primary challenge in designing a network of conservation lands is to create and foster a balance between human needs and uses of the land and environmental sustainability.

Resource management is in large part a social science. Adaptive environmental assessment (Holling, 1978) has been suggested as a useful process, because it takes into account all relevant social, economic, and environmental considerations, addressing conflicts directly and developing a framework for evaluating tradeoffs, rather than attempting to write prescriptions for problem solving. The process involves scientists, resource managers, policy analysts, and decision makers interactively in the shared responsibility for the development of a simulation model of the system to be managed. An assessment program should also be extended to monitor achievement of goals. An ecological and land-use history might also be used to understand the processes that link resources dynamically in the landscape.

In principle, Congress functions as the ultimate coordinator for federal land acquisitions, through its appropriation bills. But the existing mechanisms are inadequate for dealing with cross-agency regional, national, and even global issues, and the transfer of ecological knowledge and technology is slow. An interagency process is needed to coordinate federal land acquisition, promote and facilitate collection and transfer of information, and develop a long-term plan and strategy for land acquisition. Such cooperation is a challenge that will require institutional innovation.

# 6

# Nonprofit Organizations

During the past 2 decades, several national nonprofit organizations have assumed a prominent role in U.S. land protection. Those organizations have added flexibility to the conservation process, expanded the reach of federal dollars, and added new dimensions to the formats of land-protection transactions. One of the major nonprofit organizations, for example, summarized the types of projects in which it might become involved as

> multi-parcel assemblages; complex transactions that require intervention by a nongovernmental entity, such as those in which there is a need to liquidate a corporation; instances where agency funding does not coincide in time with landowner needs and unanticipated opportunities in the market that require prompt private action (Trust for Public Land, 1991).

In addition to the large organizations, a huge network of state and local land trusts has a major role in project identification and is a prime source of interim financing and public support. Some of the largest of those trusts also perform the same functions that national conservancies do. The best current estimates are that there are some 900 local and regional trusts nationwide (The Land Trust Alliance, 1991). For the most part, those organizations reflect an awareness of local and regional concerns and offer practical understanding of resources and of potential economic effects of acquisitions on the communities in which they operate.

Each nonprofit organization has its own criteria for selecting potential land acquisitions. The American Land Conservancy, the Conservation Fund, and the Trust for Public Land, for example, use criteria that largely reflect federal criteria (H. Burgess, pers. comm., American Land Conservancy, Feb. 13, 1991; The P. Noonan, pers. comm., Conservation Fund, Jan., 1991). But the criteria used by The Nature Conservancy (TNC), one of the largest nonprofit organizations involved in land acquisition, focus on preserving biological diversity and ecosystems. TNC has a systematic, state-by-state nationwide selection process that results in a priority site list. If a federal, state, or local agency approaches TNC for assistance in acquiring a property, TNC compares the site with the priority list. TNC becomes involved only if a site is on its list, although it will investigate a site not on it.

TNC recently launched a campaign called "Last Great Places: An Alliance for People and the Environment," to protect entire ecosystems and targeted 12 areas in the United States and Latin America for protection (Sawhill, 1991a). These areas range from 12,000 to 1,000,000 acres and have been judged as representing ecologically salvageable landscapes with functioning but endangered ecosystems that contain rare species. All 12 sites have a core natural zone of critical habitat and a surrounding buffer zone in which land-use practices might affect the zone. According to the president of TNC, the initiative fully recognizes the rights of residents in these areas and will attempt to design stewardship practices that are compatible with preservation of biological values and human interests.

Nonprofit organizations that operate on a local level often are formed for specific purposes. For example, the French and Pickering Trust was created in 1967 when a farm adjacent to French Creek in Pennsylvania was slated for development. After preventing development of that 100-acre parcel, the trust purchased other parcels in the same area for conservation. The trust eventually became a permanent organization with local goals (Morris, 1982).

To develop further the acquisition criteria used by private groups, somewhat more extended discussions of the work of three groups--the National Fish and Wildlife Foundation (NFWF), Ducks Unlimited (DU), and TNC—are presented below.

## THE NATIONAL FISH AND WILDLIFE FOUNDATION

NFWF was established by Congress as a private, nonprofit foundation to support the U.S. Fish and Wildlife Service (USFWS) and related activities; its objective is to be a catalyst for cooperative partnerships for habitat conservation. It is an operating conservation organization that raises money and gives grants. Private and state funds raised are matched by Congress and are spent on land acquisition, research programs, education, endangered species recovery, restoration of degraded habitat, and limited policy work. The policy work consists of political analyses to assist USFWS and other agencies and nonprofit organizations in working together to achieve substantive conservation goals. NFWF-acquired property rarely is held but is given to other public or private entities. The foundation has a peer-review process for evaluating projects and may agree to invest in projects if funds are matched. Key criteria for funding projects are whether they involve multiple partners and conservation values.

A top priority for the organization has been to help achieve the goals of the North American Waterfowl Management Plan (NAWMP). NAWMP is a plan to restore North American waterfowl to peak 1970 levels, largely through enhancement and protection of critical wetlands habitat. NFWF helped to organize joint-venture steering committees to develop site-specific priorities and help find funds to meet identified goals. A new group was established, the North American Wetlands Conservation Council (NAWCC), composed of nine entities. NAWCC reviews proposals for matching funds for acquisition and restoration and makes recommendations to the Migratory Bird Conservation Commission, which makes final funding decisions (see Table 6-1 for a list of NAWCC wetlands projects).

NFWF is downscaling its involvement in wetlands and shifting priorities; less focus will be put on land acquisition. A major priority for the organization will be the neotropical migratory bird conservation program and restoration of fishery resources through protection of coastal habitat, rivers, and estuaries. Management will be emphasized over acquisition, and outreach to private landowners will be conducted. According to former NFWF Director Chip Collins, land acquisition remains strategically important, but the incentives for proper management practices

TABLE 6-1 North American Wetlands Conservation Projects

| Project | State | Total Cost | Nonfederal Partner Cost | NAWCA Request | Partners |
|---|---|---|---|---|---|
| ACE Basin NWR—Year 1 | SC | 580,200 | 15,000 | 326,400 | USFWS; TNC; SC Wildlife and Marine Resources Dept.; DU, Inc. |
| Bayou Penchant | LA | 1,520,059 | 759,809 | 560,250 | TNC; Mike St. Martin; H-Bar-H Ranch; LA Dept. of Wildlife and Fisheries; NFWF; Dow, USA |
| Big Stone NWR | MN | 648,570 | 330,770 | 317,800 | DU, Inc. |
| Caddo Lake | TX | 3,854,674 | 1,928,337 | 1,924,837 | Texas Parks and Wildlife; Texas Nature Conservancy; Texas General Land Office; USFWS |
| California Central Valley | CA | 3,823,850 | 2,531,750 | 1,280,000 | Wildlife Conservation Board; CA Dept. of Fish and Game; CA Waterfowl Association |
| Cameron Creole | LA | 989,100 | 258,100 | 250,000 | LA Dept. of Natural Resources; Miami Corporation |
| Cheehaw-Combahee Reserve | SC | 8,021,900 | 5,489,928 | 2,275,972 | DU Foundation; DU, Inc.; TNC; USFWS; NFWF; SC Wildlife and Marine Resources Dept.; Cheeha-Combahee Plantation, Inc. |

# NONPROFIT ORGANIZATIONS 143

| | | | | | |
|---|---|---|---|---|---|
| Cheyenne Bottoms | KS | 8,472,828 | 4,212,828 | 2,500,000 | KS Dept. of Wildlife and Parks; USFWS; TNC; DU, Inc.; NFWF; DU, Inc.; Western Hemisphere Shorebird Reserve Network |
| Cobscook Bay | ME | 1,905,000 | 1,355,000 | 550,000 | Land for Maine Future Board; private individuals |
| Columbia Marsh | SD | 64,200 | 32,100 | 32,100 | DU, Inc. |
| Coon Island | MO | 1,912,775 | 1,275,275 | 637,500 | MO Dept. of Conservation; DU, Inc. |
| Consumnes River | CA | 692,900 | 352,900 | 340,000 | TNC; DU; DU, Inc.; BLM |
| Crow Island Wetland Initiative | MI | 553,164 | 333,164 | 175,000 | MI Dept. of Natural Resources; DU, Inc.; Waterfowl USA, Interlakes Chapter; Waterfowl USA, Saginaw Valley Chapter; New Haven Foundry; MI Duck Hunters Association |
| Deep Fork River III | OK | 514,351 | 259,351 | 250,000 | KS Power and Light Co.; OK Dept. of Wildlife Conservation; Deep Fork Coalition; Helmrich and Payne; County Commission |
| Eagle Bluffs | MO | 10,508,289 | 7,093,821 | 640,000 | MO Dept. of Conservation; DU, Inc.; City of Columbia; University of MO; Army Corps of Engineers |
| Freitas Unit I | CA | 502,100 | 0 | 245,800 | USFWS |

TABLE 6-1 (continued)

| Project | State | Total Cost | Nonfederal Partner Cost | NAWCA Request | Partners |
|---|---|---|---|---|---|
| Heron Lake | MN | 1,200,000 | 600,000 | 600,000 | MN Dept. of Natural Resources; Legislative Commission on Minnesota Resources |
| Heron Lake II | MN | 3,492,000 | 1,422,000 | 1,420,000 | MN Dept. of Natural Resources; Legislative Commission on Minnesota Resources; DU, Inc.; TNC; N. Heron Lake Game Producers Association; Middle Des Moines Watershed District; MPCA; USFWS |
| Horns Bluff | TN | 459,323 | 372,100 | 56,223 | TN Wildlife Resources Agency; DU, Inc.; TNC; TN Dept. of Conservation; Agricultural Stabilization and Conservation Service; Soil Conservation Service |
| Howard Slough | CA | 240,000 | 121,500 | 118,500 | CA Dept. of Fish and Game; CA Waterfowl Association; National Audubon Society; TNC |
| Iowa Prairie Pothole | MN | 325,000 | 75,000 | 75,000 | Iowa Dept. of Natural Resources; DU, Inc.; USFWS |

## NONPROFIT ORGANIZATIONS

| | | | | | |
|---|---|---|---|---|---|
| Kelly's Slough | ND | 395,745 | 205,365 | 190,380 | DU, Inc.; Grand Fork County Wildlife Club; township boards; County Engineers; USFWS; US Air Force |
| Kelly's Slough II | ND | 546,828 | 223,253 | 214,497 | DU, Inc.; USFWS |
| Llano Seco Unit | CA | 240,000 | 121,500 | 118,500 | CA Dept. of Fish and Game; CA Waterfowl Association; National Audubon Society; TNC |
| Mad Island Marsh | TX | 1,255,500 | 730,500 | 425,000 | TNC; DU, Inc.; NFWF |
| Maple River | MI | 160,000 | 125,000 | 35,000 | MI Dept. of Natural Resources; DU, Inc. |
| Maurice River Basin | NJ | 3,913,500 | 2,613,500 | 1,300,000 | State of NJ–Green Acres; Natural Lands Trust; TNC; Waterfowl Stamp Program |
| Metzger's Marsh | OH | 1,450,000 | 729,500 | 700,000 | Ohio Division of Wildlife; DU, Inc.; Herb Metzger; Maumee Audubon; Lake Erie Wildfowlers; Decoy Carvers/Collectors; Bluffton Sports Club; Wolf Creek Sports Club |
| Mustin Bottoms I | TN | 2,657,700 | 1,675,600 | 882,100 | TWRA; TNC; private landowner |
| Nanticoke River Wetlands | DE | 659,079 | 374,079 | 285,000 | DE Division of Fish and Wildlife; TNC |
| NE Montana Prairie Pothole | MT | 917,500 | 142,500 | 100,000 | DU, Inc.; MT Dept. of Fish and Wildlife; TNC; USFWS; NFWF |

**TABLE 6-1** (continued)

| Project | State | Total Cost | Nonfederal Partner Cost | NAWCA Request | Partners |
|---|---|---|---|---|---|
| North Landing River | VA | 1,336,466 | 709,862 | 626,604 | City of Chesapeake; VA Dept. of Conservation and Recreation; DU, Inc. |
| North Landing River II | VA | 4,204,900 | 2,245,000 | 1,755,400 | Day Companies; TNC; VA Dept. of Conservation and Recreation; Army Corps of Engineers; EPA/VA |
| Oklahoma Playa Lakes | OK | 119,000 | 69,000 | 50,000 | OK Dept. of Wildlife Conservation; Phillips Petroleum Co. |
| Pickerel Creek II | OH | 1,105,000 | 655,000 | 450,000 | Ohio Division of Wildlife; DU, Inc.; Ohio Plan Clubs; Maumee Valley Audubon |
| Playa Lakes | KS | 200,000 | 100,000 | 100,000 | KS Dept. of Wildlife and Parks; DU, Inc.; Phillips Petroleum |
| Private Land Restoration | MN | 141,000 | 79,000 | 62,000 | MN Waterfowl Association; MN Board of Soil and Water Resources; MN Dept. of Natural Resources |

| | | | | |
|---|---|---|---|---|
| Rim Restoration | MN | 1,522,905 | 1,356,930 | 165,000 | MN Board of Soil and Water Resources; 26 soil and water conservation districts; Pheasants Forever, Inc.; MN Dept. of Natural Resources; sportsmen's groups; USFWS; Soil Conservation Service |
| Salem River | NJ | 4,080,000 | 2,080,000 | 2,000,000 | NJ Division of Fish, Game, and Wildlife; TNC; NJ Waterfowl Stamp Program; DU, Inc.; Natural Lands Trust; NJ Dept. of Agriculture; Salem County Agriculture Development Board |
| San Carlos Wetlands | AZ | 58,650 | 29,325 | 29,325 | San Carlos Apache Tribe |
| Schindler WPA | NE | 1,410,000 | 60,000 | 60,000 | USFWS; DU, Inc. |
| Sediment Diversions | LA | 147,000 | 85,000 | 32,000 | LA Dept. of Wildlife and Fisheries; LA Dept. of Natural Resources; Texaco; Chevron; NFWF; USFWS |
| Sediment Diversions II | LA | 160,000 | 100,000 | 60,000 | LA Dept. of Natural Resources; LA Dept. of Wildlife and Fisheries |
| Swan Lake | MN | 1,360,000 | 685,000 | 675,000 | MN Dept. of Natural Resources; Legislative Commission on Minnesota Resources |

TABLE 6-1 (Continued)

| Project | State | Total Cost | Nonfederal Partner Cost | NAWCA Request | Partners |
|---|---|---|---|---|---|
| Swan Lake II | MN | 1,116,850 | 553,425 | 553,425 | MN Dept. of Natural Resources; Legislative Commission on Minnesota Resources; Nicollet Conservation Club; Mankato State University |
| Ten Mile Pond | MO | 3,007,375 | 2,668,625 | 338,750 | MO Dept. of Conservation; DU/Marsh |
| Thielke Lake | MN | 550,000 | 275,000 | 275,000 | MN Dept. of Natural Resources; DU, Inc. |
| Thomas M. Goodwin | FL | 300,000 | 200,000 | 100,000 | FL Game and Freshwater Fish Commission; DU, Inc. |
| Thomas M. Goodwin II | FL | 253,500 | 169,000 | 84,500 | FL Game and Freshwater Fish Commission; DU, Inc. |
| Thomas M. Goodwin III | FL | 292,000 | 154,000 | 138,0000 | FL Game and Freshwater Fish Commission; DU, Inc.; St. Johns River Water Management District |
| Thousand Springs/ Chilly Slough II | ID | 277,900 | 125,600 | 125,000 | TNC; DU, Inc.; Idaho Fish and Game; BLM |
| Traction Ranch | CA | 581,972 | 296,402 | 285,570 | CA Dept. of Fish and Game |

| | | | | |
|---|---|---|---|---|
| Trice Wetlands | AR | 143,710 | 75,710 | 68,000 TNC/DU, Inc.; AR Natural Heritage Commission; AR Game and Fish Commission |
| Uthlein WPA | MN | 343,000 | 192,000 | 51,000 DU, Inc.; NFWF |
| Upper Butte Sink | CA | 652,619 | 335,508 | 317,111 CA Dept. of Game and Fish; DU, Inc. |
| Upper Lightening Lake | MN | 102,000 | 51,000 | 51,000 MN Dept. of Natural Resources |
| Warner Valley | OR | 330,500 | 60,000 | 215,000 TNC; DU, Inc.; Izaak Walton League of America |
| Wetland Development on Private Lands | SD | 1,234,300 | 569,300 | 445,000 DU, Inc.; SD Association of Conservation Districts; 35 SD conservation districts; Rosebud Sioux Tribe; Sisseton-Wahpeton Sioux Tribe; private landowners; USFWS; Soil Conservation Service; Bureau of Indian Affairs |
| Wetland Restoration and Private Lands | MS | 1,691,491 | 664,671 | 474,000 DU, Inc.; USFWS; Soil Conservation Service; MS Dept. of Wildlife, Fisheries, and Parks; LA Dept. of Wildlife and Fisheries; Texas Parks and Wildlife Dept.; Transco; Arkla Energy; SONAT Foundation; Texas Gas |

TABLE 6-1 (continued)

| Project | State | Total Cost | Nonfederal Partner Cost | NAWCA Request | Partners |
|---|---|---|---|---|---|
| White Lake | TN | 1,058,900 | 899,400 | 128,500 | TN Wildlife Resources Agency; Soil Conservation Service; DU, Inc.; TDOC; ASCS |
| White Lake II | TN | 624,929 | 451,804 | 173,125 | TWRA; DU, Inc. |
| Wigwam Bay | MI | 221,000 | 171,000 | 50,000 | MI Dept. of Natural Resources; DU, Inc. |
| Williams-Asche Wetlands | DE | 1,990,000 | 819,560 | 800,000 | Longwood Foundation; TNC Campaign for the DE; Crystal Trust; Welfare Foundation; Delmarva Power Co.; Grace V. Asche; HTS Environmental Services (in kind); TNC Land Preservation Fund |

Source: National Waterfowl and Wetlands Office, USFWS, January 27, 1993.

must be addressed. Migratory bird conservation programs have met resistance with the exception of duck conservation, and USFS will be a critical agency to work with in this endeavor, because forests provide most of the habitat for songbirds.

An example of a project in which NFWF played an instrumental role is "ACE Basin" in South Carolina, which is one of the last unprotected and undeveloped coastal estuarine areas. It consists of 300,000 acres and encompasses the convergence of the Ashepoo, Combahee, and Edisto rivers and more than five ecosystems from the barrier islands to the bottomland hardwood forests. The area also includes freshwater impoundments, provides habitats for many rare and endangered species, and has high concentrations of nesting southern bald eagles and osprey. Several entities cooperated in this project, including USFWS, the National Oceanographic and Atmospheric Administration, TNC, and the state. The conservation strategy was developed by a task force, and conservation easements were donated by private landowners. A similar project was conducted on the Cache River in Illinois.

The largest project of this type was the Parrot Ranch near Chico, California. The ranch was an old Spanish land grant still owned by the original family. The land is a significant property in the Central Valley, in part because it has senior water rights. In this effort, a management committee was formed by six different entities.

## DUCKS UNLIMITED

Ducks Unlimited (DU) was organized in 1937 to preserve and protect wetlands habitat for the benefit of waterfowl and wetlands wildlife. DU subscribes to the proposition that loss of habitat is the most severe problem confronting wildlife and plant resources worldwide.

DU's early efforts focused on raising funds in the United States for use in Canada, where 5,049 projects involving 2,194,387 acres have been completed. Since 1974, when DU began initiatives in Mexico, 99,231 acres have been protected at 82 different project sites. In 1984, DU began a domestic conservation program in recognition of the importance of migration rest-stop and wintering sites in the United States. Since then, 495 projects in all the states have been completed, encompassing nearly 450,000 acres of wetland and upland habitat.

DU participates in the implementation of NAWMP, which involves grassroots partnerships to protect private and public lands. Table 6-1 is a list of current NAWMP projects; DU is a partner in most of them.

Each potential site identified by DU is visited by a DU biologist. For each habitat project considered, an evaluation identifies the following:

- Type of habitat and acreage involved;
- Waterfowl and other wetland wildlife species expected to benefit;
- Type of development and enhancement activities;
- Long-term management objectives as well as operations and management techniques to be used;
- Biological considerations, e.g., average water depth to be maintained, plant types and distribution, proximity of nesting cover to the wetland, anticipated nesting success, surrounding land use, waterfowl abundance, probability of disease outbreaks, uniqueness of habitat, importance to endangered species, and frequency of hunting in the area;
- Public relations values, including proximity to population centers, importance to members, and potential for fundraising;
- Other considerations, such as cost effectiveness, conformance with NAWMP, and cooperators in the project;
- Estimated cost, sources of funding, type of acquisition, and term of agreement if fee-simple acquisition is not involved.

The prominent acquisition devices used by DU are the purchase of fee-simple title and the purchase or donation of an easement. However, DU is not committed to the exclusive use of any particular device. According to DU, fee-simple title acquisition of land should be used to purchase the most environmentally sensitive areas where no other comparable habitat remains and to purchase core areas or key tracts of land in an otherwise larger area of enhancement and preservation.

DU recognizes that neither the public nor private sector has sufficient funds to purchase all the habitat deemed essential. Furthermore, continuing expenses are associated with managing the land once it is acquired. For these and other reasons, DU uses conservation easements as well as outright purchase.

DU invests in the enhancement and preservation of critical habitats owned by others through matching aid to restore state habitats and private land programs. In these cases, DU requires a site-specific agreement or other appropriate contract.

## THE NATURE CONSERVANCY

TNC owns and manages more than 1,600 nature preserves, the largest of which is the 500-square-mile Apecheria ranch in New Mexico. TNC has the largest private conservation program in the country with $600 million in assets, and it owns about $67 million in lands for trade. Of contributed funds during 1990, 78.7% came from individuals, 11.6% from corporations, and 9.7% from foundations. In 1990, support and revenue totaled $137,734,000 (TNC, 1990).

Criteria for land-acquisition priorities are detailed in TNC's *Preserve Selection and Design Manual* (PS&D) (see Appendix D). TNC criteria are internally consistent, workable, and scientifically defensible criteria for maintaining biological diversity.

TNC identifies occurrences of elements of diversity (referred to as "element occurrences" or "EOs"), which are any type of biological or ecological entity (e.g., a species or community) in a geographic area. EOs are mapped and ranked as the basis for preserve selection and design. Sites also are ranked for other values—other uses or benefits they might have besides their potential contribution to conservation. Those values include ecological service functions, such as aquifer recharge and erosion prevention, and other benefits such as recreation, aesthetic enjoyment, and historic and archeological significance. This helps to identify potential partnerships and increase the feasibility of implementation (R. Jenkins, pers. comm., TNC, Jan. 8, 1991).

TNC's heritage programs contain the largest and most extensive data base for rare species, communities, and ecosystems in the world. This could be a valuable source of biological information for the four federal agencies that are the subject of this study.

### Minimal Size of Preserves

The PS&D states that preserves may lose species over time, particularly if they are small but that the causes of species loss are not well understood, and the probability of loss differ among taxa. For example, many plant species (e.g., prairie flora) have existed over a long time on small plots. Practicality is often the overriding issue: A big preserve might be best, but it also can be expensive or nonexistent because of habitat loss and fragmentation.

Although TNC uses the term "minimum viable size of preserves," it is generally understood that size varies with the EOs targeted for preservation. A hectare may be sufficient to preserve a self-fertilizing, long-lived plant, but 500,000 hectares in a fragmented landscape might be insufficient to preserve a population of a large vertebrate predator. Nonetheless, it might be possible to classify different types of EOs and provide some general guidelines for optimal to minimal reserve sizes.

## Minimal Population Size

TNC recognizes that small populations are more subject to extinction than large populations because of genetic and environmental reasons (e.g., demographics, environmental variation, and natural catastrophes) but states that minimal population size cannot realistically be determined by analyzing any of these factors. The PS&D concludes that each preserve is an experiment in minimal population size, and the reserve designer should try to prescribe boundaries and management practices that minimize extinction. TNC tries to ensure that rare EOs are preserved in enough places that the likelihood of all becoming extinct is small.

The committee believes that more precise and relevant terminology could be developed for population and genetics. In addition, it might be possible to develop more general guidelines than those mentioned above by drawing upon knowledge of extinction, evolutionary trends, and the genetic structure of populations. Even if generalizations are not possible, a list of factors could be provided that could guide the preserve designer more objectively. Such factors include genetic structure, population size, effective breeding population size, breeding structure, longevity, net reproductive rate, degree of niche specialization, stability/predictability of local climate, probability of natural catastrophes, and landscape information. Available information on those factors for EOs designated for preservation could be collected. Although much of this information is not available for most species, an ordinal ranking scale could be developed, as is done for EOs.

The committee understands why TNC might be reluctant to adopt such a system. If the ranking suggests that a preserve might be too small to sustain an EO for a reasonable length of time, the mission of TNC directs it to err on the conservative side—even if the best scientific information suggests a low chance of success, TNC might attempt preservation, because extinction is permanent, and science often is wrong.

## Preserve Configuration and Justification

The PS&D states that preserve-design theory suggests that preserves should be as round as possible (to minimize edge) and connected by corridors to facilitate migration. But preserve shape and juxtaposition are difficult to control, because of ownership patterns and past habitat disturbance. If an EO does become extinct on a preserve, TNC will consider reintroduction from elsewhere.

Usually, tracts available for preserves are much smaller than desirable preserve size, but accumulating several unconnected tracts containing rare EO's might reduce the probability of extinction, because if local extinctions occur, reintroductions are possible.

Opinion on the value of corridors varies. The concept is not well defined, according to a member of the TNC board of directors (Stolzenburg, 1991), and corridors are considered an inadequate substitute for suitable habitat. Simberloff and Cox (1987) point out also that corridors can provide access for parasites, predators, pests, fire, and poachers.

Noss and Harris (1986) are strong advocates of corridors, and Stolzenburg (1991) described three studies that demonstrate the efficacy of corridors for maintaining populations: Harper in Brazil found that corridors are essential for maintaining antbirds in patches of jungle, Bennet in Australia discovered that corridors provide transportation and a conduit for gene exchange, and Merriam in Canada found that woodlots connected by wooded fencerows demonstrate a continual process of extinction and recolonization by small mammals and birds (Stolzenburg, 1991).

Some TNC conservation efforts have incorporated corridors. Merrill Lynch (TNC, North Carolina) designed the corridor in the 30,000-acre Pinhook Swamp reserve in northern Florida that was a cooperative effort between TNC and the Forest Service. Lynch is planning a multicorridored project—436,000 acres of the Alligator River Wildlife Refuge (in collaboration with the Fish and Wildlife Service, TNC, and The Conservation Fund)—targeted for black bear and red wolf protection (Stolzenburg, 1991). And TNC's "The Last Great Places" initiative clearly is directed at saving landscapes with core habitat surrounded by buffer zones.

### Element and Site Stewardship

TNC insists that proper stewardship requires highly specific information about the EO and site on which it occurs. Several years might be needed to collect the natural history of the EO and the possible constraints on managing for its survival. The PS&D manual, however, does not help the field staff set priorities for information to be collected based upon scientific data. The manual does not provide a review of pertinent ecological or conservation literature from either an empirical or theoretical perspective.

This design criterion of the TNC acquisition strategy shows an attentiveness to practicality. PS&D briefly dismisses most theoretical issues concerning preserve design and focuses on what site-specific information is needed and can be obtained. For example, for an endangered plant, what are the pollinators, seed-dispersal agents, soil type, and habitat affinity? Are there threats from neighboring lands and use of herbicides?

This type of detective work is absolutely necessary and, based on the history of TNC preservation efforts, successful. Site-specific data from field surveys is one key to TNC success. It is important, however, that data gatherers be instructed in the empirical and theoretical literature that clarifies how to monitor and measure the ecological condition of a particular site. The PS&D manual and its successor documents might be one way to accomplish this.

As part of the "Last Great Places" bioreserve initiative, TNC is giving more attention to landscape-level considerations. This requires more emphasis on the integration of human needs into conservation efforts through sustainable development schemes and cooperative management strategies for multiuse landscapes. In the future, project selection process probably will favor sites that are or can be included in landscape complexes over isolated sites of more dubious long-term viability (R. Jenkins, pers. comm., TNC, Jan. 8, 1991).

# 7

# Techniques and Tools of Acquisition

The acquisition of land in its most common form is a familiar process. Land is identified and appraised, a price is negotiated, the escrow is established, and the title is transferred. Those steps can be completed by an agency of the federal government, by a nonprofit corporation, or by a nonprofit organization in cooperation with one or more federal agencies. Land also can be acquired by joint ventures between federal and state or local agencies.

The range of options and methods available for land acquisition often are grossly underestimated. Those include the reservation of future interests in the management or ownership of property, reversion of ownership if a management agreement is violated, the right of entry to enforce reversion, and conveyance of partial interest. Common-law techniques include easements and covenants.

Other techniques that have resulted from statutes, estate planning, and creativity include zoning, leases, purchase and lease- or sell-back arrangements, dedications, management agreements, bargain and installment sales, purchase options, rights of first refusal, covenants prohibiting sale, transfers in trust, transferable development rights, statutory easements, and scenic easements (Hoose, 1981; Barrett and Livermore, 1983; Brumbach and Brumbach, 1988). In the past 15 years, at least 128 legal articles have been published on less-than-fee acquisition and conservation techniques. Many of those deal with the use of conservation easements and similar techniques in special contexts—highway scenic easements (Cunningham, 1967), open-space laws (Hoffman,

1989), preservation of agricultural land (Hamilton, 1985), historic preservation and conservation (Wilson and Winkler, 1971; Netherton, 1979), and tax policies that have encouraged modern conservation easements (Madden, 1983). The choice of method used depends on the goals to be achieved; for example, partial interests, such as the transfer of development rights or acquisition of easements restricting land uses, might be sufficient for preserving open spaces, forests, or farmlands but might be inadequate when public access is desired.

This chapter discusses a few of the more prominent techniques and procedures for federal land acquisition, and provides some examples of the participation of nonprofit organizations in the process.

## CONSERVATION EASEMENTS

Most states have statutes authorizing the creation of conservation easements, several of which are variations on the Uniform Conservation Easement Act. One of the most comprehensive statutes is the California Conservation Easements Act of 1979, which, although not necessarily typical, is presented as an illustration. Under that act, a conservation easement is an interest in property that is binding upon successive owners with the purpose of retaining land "predominantly in its natural, scenic, historical, agricultural, forested, or open-space condition" (Cal. Civ. Code Section 815.1). That act abolishes technical distinctions between easements, covenants, and conditions. A conservation easement is freely transferable, is binding on the owner of the restricted land, and avoids the classical limitation that the easement must benefit particular lands. A conservation easement is perpetual and is enforceable by injunction at the behest of the grantor or holder of the easement. In addition to injunctive relief, "money damages also will lie, not only for the costs of restoration but for the loss of scenic, aesthetic or environmental values" (Barrett and Livermore, 1983).

To ensure that easements are used for conservation purposes, the only entities that can hold and acquire them are government entities and tax-exempt, nonprofit organizations that have as a primary purpose retaining the land for the reasons above. This limitation has origins in the federal tax laws, which dramatically increased the use of conservation easements once they were permitted to be deducted as charitable contributions in the late 1960s (Barrett and Livermore, 1983). The federal government and its agencies are not among the entities that can purchase conservation easements in California; therefore, federal agencies are obliged to construct easements enforceable under common law.

Approximately 2 million acres of U.S. land is subject to conservation easements held by approximately 500 nonprofit organizations and government entities (Brumbach and Brumbach, 1988). Conservation easements have been used with varying success. The U.S. Fish and Wildlife Service (USFWS) has used easements since the 1950s to preserve waterfowl breeding habitats in Minnesota and the Dakotas. USFWS purchased perpetual easements to prohibit burning, draining, and filling wetlands for more than 1,100,000 acres. Purchase of that land would have been prohibitively expensive, but the easement approach focused on essential property rights for saving crucial waterfowl habitat and made it possible for USFWS to protect much more habitat than it could have otherwise (Barrett and Livermore, 1983). The gains from those less-than-fee acquisitions, however, must be measured against the costs, one of which is diminished ability to enforce the restrictions and ensure compliance with the protective conditions (GAO, 1989).

Another experience with conservation easements is that of the National Park Service (NPS) in its program to build scenic parkways through large wilderness areas during the 1930s. In part to keep costs down and in part because the primary public use of the adjacent land was to be the observation of scenic beauty, NPS purchased some of the lands in fee and used easements to obtain the rest: for each 100 acres purchased, conservation easements were used for another 50 acres per mile of parkway. Easements were purchased for 4,500 acres along the Natchez Trace Parkway in Alabama, Mississippi, and Tennessee, and on 1,200 acres along the Blue Ridge Parkway in Virginia and North Carolina. The easements prevented cutting trees, erecting billboards or signs, dumping trash, and engaging in any other activity that effectively changed the existing use of land (Matuszeski, 1966). NPS abandoned further easement acquisition for several reasons, including problems resulting from state agents negotiating transactions while the federal government remained responsible for enforcement:

> It has been suggested that this procedure resulted in the landowners not being fully apprised of what rights they were yielding, since the state agent had to

concern himself only with getting the landowner to sign his name to the agreement. Yet it was the federal enforcing agents who were to receive full blame when Farmer Jones found out that he couldn't clear another forty. It was extremely difficult to explain to him that he had sold a bundle of rights in his property to the federal government when he still held the deed and used the land daily as he had for years previously. Problems were compounded when a second generation of owners came along—the sons who had not signed any agreement and did not feel bound by it, even if they knew of it. Meanwhile the costs of patrol for easement enforcement had increased substantially, while violations of the agreement became more frequent. Perhaps most crucial were the decisions of local and U.S. District courts in the area, which consistently refused to grant the full injunctive relief requested by the federal government. In the face of all these hostile factors, the NPS stopped the purchase of further easements and converted to a fee purchase program on both parkways (Matuszeki, 1966).

These and other experiences show the mixed motivations and results that tug and pull at the popularity of the conservation easement. The device can be attractive for land-acquisition agencies, because it avoids condemnation and allows leveraging of scarce dollars. Easements often originate from philanthropic motives of landowners (Ward and Benfield, 1989), which might aid enforcement, at least in the short term. Among the problems resulting in ineffective or failed conservation easements are

- Complexity of initial drafting and excessive ambiguity;
- Rigidity of easement provisions that fail to tailor easement to property characteristics and account for reasonable retained uses;
- Failure to monitor easement regularly and thoroughly; failure to plan for costs of monitoring;
- Failure to establish ongoing relationships with owners of encumbered property, including new owners;
- Inappropriate use of easements (e.g., when significant public access and management is desirable);
- Failure to enforce terms of easements that can reinforce patterns of laxity;
- Viewing the landowner-agency relationship as adversarial.

Because of these and other difficulties with easements, Reitze (1974) suggests that conservation easements are "most useful when passive uses not involving physical occupation by the public are contemplated."

Drafting conservation easements to ensure the desired protection is a challenge. The commitment to protection is reinforced by giving the holder of the easement various rights to preserve, protect, and enhance the natural and ecological values of the property, to enter upon the property at reasonable times to enforce the rights granted and make studies, and to seek an injunction and restoration if the property is damaged contrary to the terms of the easement.

Interested third parties, such as conservation or environmental groups, generally are not allowed to enforce easements they do not formally hold. Therefore, enforcement cannot be accomplished by private citizens bringing lawsuits, as they can under many of the federal environmental statutes.

## TRANSFERABLE DEVELOPMENT RIGHTS

Transferable development rights (TDRs) are another recent addition to the types of interests in property that might help to preserve the current state of a parcel. An offshoot of zoning restrictions, TDRs permit a landowner to sell the right to develop a property that has been foreclosed by regulation to another individual who owns land that can be developed; the principle is that higher-density development at the receiving site will be offset by the open space at the transferring site.

A variety of TDR programs have been analyzed in the legal literature on land use. TDR programs

generally are implemented to channel development away from environmentally sensitive land areas and toward designated growth areas. The programs allocate permits for development efficiently when communities desire to limit the total amount of development. Land use programs incorporating TDRs generally designate some land as preservation areas, where little or no development is allowed, and other land as growth areas, suitable for high density residential or commercial development. The local land-use regulatory authority grants TDRs to property owners in the preservation area, which they can sell or transfer to other tracts. Once property owners in the preservation area sell their TDRs, they must register a conservation easement on their property deeds permanently restricting the development of the land. The only other means to obtain a high degree of permanence for land-conservation purposes—acquiring the land—is much more expensive, and therefore much less attractive to communities than TDR programs (Tripp and Dudek, 1989).

Use of TDRs avoids challenges that might be made to more straightforward zoning or regulatory restrictions. But it does so partly by making concessions. To use TDRs effectively, state or local governments must have meaningful land-use controls; otherwise the development right has no economic value. Successful TDR programs (including pollution rights) depend on many variables, including creating evasion-proof trading schemes, establishing rights that have economic value, and allowing transfers with minimal transaction costs (Tripp and Dudek, 1989). Under those criteria, emissions trading plans designed to reduce air pollution in Los Angeles and a pollution rights trading scheme to reduce pollution in the Fox River, Wisconsin, are failures.

On the other hand, the New Jersey Pinelands TDR program is a success. That program imposes conservation easements on properties in preservation and agricultural production areas and pays off the owners with TDRs that can be sold to developers, who use them to increase permitted density in building areas. Located in southeastern New Jersey, the New Jersey Pinelands is a national reserve that contains approximately 1 million acres of forests, wetlands, creeks, and rivers. Tripp and Dudek (1989) describe the TDR program as the "most ambitious, innovative, and geographically extensive one" in the country. The success of the program can be attributed to a variety of factors, including the competence of the commission staff that manages the program, ensured economic value for development rights, and the evasion-proof nature of the trading scheme. Tripp and Dudek also underscore the importance of specifying clearly the resource-protection objectives.

## DEDICATION

Dedication is "the placement of a natural area into a legally established system of nature preserves, whose member properties are protected by strong statutory language against condemnation or conversion to a different use. The preserve system is administered and usually managed by a state agency" (Hoose, 1981). Twelve states, mostly in the midwest, have laws for dedication arrangements. Landowners usually can dedicate specific interests or full-fee title in property. Hoose uses the example of an owner dedicating the rights to cut the trees in an old-growth stand while maintaining the rights to live on or transfer the property.

Public trusts are the common-law version of statutory dedication. The public trust doctrine holds generally that some types of natural resources (such as navigable waters and wetlands) should be held in trust for the benefit of the public (Sax, 1970; Rodgers, 1986). Those resources are protected by trust against unfair dealing and dissipation, suggesting the need for procedural correctness and care if reallocation is considered. Private property rights acquired as part of trust resources are subject to prior dedications or overriding trusts that protect other uses. For example, the rights of Los Angeles to withdraw water from Mono Lake in northern California are subject to a pre-existing obligation to allow sufficient flows to the lake to maintain its populations of migrating birds (*Nat'l Audubon Society* v. *Superior Court of Alpine County*, 189 Cal. Rptr. 346, 658 F.2d 709 (1983)).

## REGULATION

Direct regulation is an obvious technique that can obviate the need for acquisitions. An interesting example is that of the Endangered Species Act (ESA), which envisages designations of critical habitat for endangered or threatened species. Critical habitat is defined in terms of the geographic areas occupied by the species. Landowners restricted by critical habitat definitions obviously will view them as the functional equivalent of an easement, dedication, trust, or other restriction.

To elaborate upon a prominent illustration, the original critical habitat proposals for the northern spotted owl included 11,639,195 acres in California, Oregon, and Washington, of which 3,020,529 were owned privately (FWS, 1992). Conservatively estimated, the costs of acquiring the privately owned timberland would exceed $3 billion, well in excess of the annual operating budget of the U.S. Forest Service (USFS). If critical habitat restrictions were to be imposed, millions of acres would be brought under land-use restraints, because the administrative definition of harm forbids a variety of activities that degrade the protected habitat. Concern for the vast territory covered by the critical habitat for the northern spotted owl shows that future management obligations might obviate classical distinctions between private and public ownership and that full-fee acquisition by the government of all interests necessary to achieve conservation goals is implausible.

Another example of a pervasive, if indeterminate, regulatory presence that can affect expectations and use of broad expanses of land is found in Section 302(b) of the Federal Land Policy and Management Act (FLPMA). That statute declares that, in managing the public lands, the relevant secretary "shall, by regulation or otherwise, take any action necessary to prevent unnecessary or undue degradation of the lands." Although "legislative servitudes" of this sort leave much to the imagination and to the engines of future policy choice, they have the potential to redefine traditional public and private property domains. The Bureau of Land Management (BLM), for example, faces choices of this sort:

> The FLPMA command to prevent unnecessary or undue degradation of the public lands perhaps could be stretched into a declaration that a private property interest never includes the right to interfere with important public land values. Under this interpretation, the government would hold a servitude on behalf of the public lands. This servitude would define the rights of private owners of property adjacent to the public lands and justify the imposition of restraints on the use of such private property, so long as those restraints were reasonably related to preserving collective values. Because the private party would never have "owned" the right to degrade the public lands, regulations designed to prevent degradation would not take any property (Mansfield, 1991).

Two other illustrations of incentive-changing statutory intrusions upon land-management choices are the Conservation Reserve Program (CRP) established by the 1985 Farm Bill and the Forest Legacy Program prescribed in the 1990 Farm Bill. Like the soil bank program of the 1950s, the aim of CRP was to reduce erosion—an obvious conservation purpose. It authorizes federal payments to farmers who remove erodible cropland from tillage and plant the land with cover crops or trees. The goal in 1985 was to remove 40 million acres (a maximum of 45 million acres) from tillage. Forestry interests hoped that one-eighth of this would be planted with trees. About 2.3 million acres, or 6.6 % of the total, have been planted with trees since 1985, with the remainder planted with grass and other cover.

Landowners who plant trees or cover crops on erodible cropland under CRP agree that the land will remain in that cover for at least 10 years. Landowners receive annual rental payments, and the federal government also pays for 50% of costs for some practices, such as tree planting. If the landowner returns to planting crops, all funds received must be repaid with interest plus 25% of 1 year's rental payment. The Agricultural Stabilization and Conservation Service programs of the Department of Agriculture monitor for compliance.

Trees planted will be of merchantable size after 20 or 30 years. The landowner at that time can decide whether and how the harvesting will occur and whether the land will be reforested.

Land enrolled under the CRP is under contract to the federal government. Under the 1985 CRP, no easement or other encumbrance is placed on the title to the land. The 1990 Farm Bill does provide for easements in some CRP cases. Under Section 1432, for example, properties eligible for the program including "newly established living snowfences, permanent wildlife habitat, windbreaks, shelter belts, or filterstrips devoted to trees or shrubs" are subject to an easement for their "useful life," which is defined administratively as 15 years for grass and 30 years for trees.

The 1990 Farm Bill also authorizes the Forest Legacy Program, which is designed to be a cooperative endeavor among levels of the government to identify environmentally important forest areas that are threatened by conversion to nonforest uses. The use of conservation easements to promote land protection and other conservation purposes is expected. Like the CRP, the Forest Legacy Program faces difficult issues of defining protected areas, conservation easements, and other mechanisms; fashioning incentives; and drawing lines between private entitlement and public expectation. The Forest Legacy Program is an experiment in institutional land conservation now under way.

The Fifth Amendment to the U.S. Constitution provides expressly that private property shall not be taken for public use without just compensation. That provision, along with similar measures in state constitutions, has spawned a vast body of case law and legal writing exploring whether government intrusions upon the expectations of private property owners rise to the level of a compensable taking. Indeed, this "takings" issue dominates the decisionmaking of property lawyers, managers, and public officials across the spectrum of the U.S. public and private property systems. It specifically affects decisions as to when regulatory restrictions on land uses are sufficient for achieving conservation goals and when other means of protection, such as land acquisition, are necessary.

Whatever the dimensions of the legal doctrine of takings, it provides the background for defining the boundaries between public and private expectations in lands held and acquired by the federal government. But the takings doctrine never has made every change in landowner expectation and use a constitutionally questionable issue. One observer of public land law points out that the traditional systems of property law that cut land into arbitrary pieces of 160 acres and deliberate "checkerboarding" of ownership, disclosed as a fundamental purpose the destruction of "the functioning of natural resource systems" (Sax, 1991). As the law moves in the direction of restoring and protecting natural systems, it will affect the structure of property ownership and operate under the guidance of the takings doctrine. Equally important, however, it will act through changes in use. As Sax (1991) points out, the public already owns the national forests: "The issues there are not proprietorship or compensation, but how to allocate the land between such competing demands as timber production, hydrocarbon or geothermal development, and wilderness and wildlife."

## LAND EXCHANGE

Land exchange between private parties or nonfederal public agencies and the USFS or BLM is an established means of improving land ownership patterns and administrative efficiency and achieving federal land-management objectives. It is particularly useful in areas where historical land settlement resulting from earlier federal government policies has resulted in cumbersome, fragmented ownership patterns and where nonfederal inholdings are included within large tracts of national forest or public-domain lands. In such situations, exchange of lands between a nonfederal landowner and a federal land management agency can be a means to achieve the future land ownership and management objectives of both parties without requiring large sums of money on either side of the transaction to purchase lands outright.

Land exchange is the principal means of accomplishing land ownership adjustments for the USFS. In the past 80 years, USFS has completed approximately 8,000 separate land exchanges, acquiring almost 9.5 million acres of nonfederal land in exchange for approximately 3.5 million acres of federal land. In an average year, USFS completes 147 separate exchanges, acquiring 135,000 acres from willing nonfederal landowners in exchange for 92,000 acres of federal land, with exchanged values of $102 million. Land exchanges frequently are used to acquire nonfederal lands in congressionally designated wilderness areas, national recreation areas, and wild and scenic rivers areas (*Federal Register*, Vol. 54, No. 159, 1989).

Land exchange also has been used by BLM to accomplish its land adjustment and land management objectives. An example of the importance of land exchange to BLM is illustrated by the exchanges completed in one state alone—Arizona—during fiscal years 1983-1991. In a series of land exchanges with the state and several private parties, BLM acquired 1,554,198 acres of state and private lands and conveyed 1,073,000 acres to nonfederal parties. The lands that BLM acquired included inholdings within wilderness areas and the Grand Canyon National Park, key riparian habitats, and extensive areas of native grasslands (BLM, 1991). Acquisition of these areas through land exchange has greatly enhanced BLM's ability to fulfill its conservation mission and at the same time has improved administrative efficiency by reducing the number of small tracts that BLM was responsible for managing. One of these acquisitions, since designated by Congress as the San Pedro Riparian National Conservation Area, enabled BLM to acquire title to a tract of river bottom land 33 miles long and 3 miles wide that supports 345 species of birds, 82 species of mammals, and 47 species of reptiles and amphibians, and contains two significant archeological sites (Negri, 1989).

Authority for the acquisition of lands through exchange is provided to USFS under the Weeks Law (Ch. 186, 36 Stat. 961, as amended; 16 U.S.C. 485), the General Exchange Act (Ch. 105, 42 Stat. 465; 16 U.S.C. 485, 486), the National Forest Management Act (P.L. 94-588, 90 Stat. 2949, as amended; 16 U.S.C. 516, 518, 521b), and the Federal Land Policy and Management Act (P.L. 94-579, 90 Stat. 2743, as amended; 43 U.S.C. 1701, etc.). FLPMA and other laws also provide land exchange authority to agencies of the Department of the Interior, including BLM, USFWS, and NPS. In addition to the general exchange authority provided in the laws noted above, exchange authority for specific land exchanges sometimes is provided through amendments to other authorizing or appropriations bills. This course of action has been used by Congress in recent years to expedite the completion of land exchanges that otherwise would take several years to complete following the normal process.

In recognition of the importance of land exchange as a means of securing federal land management objectives and that the "needs for land ownership adjustments and consolidation consistently outpace available funding for land purchase by the Federal Government," Congress in 1988 enacted the Federal Land Exchange Facilitation Act (P.L. 100-409, 102 Stat. 1086; 43 U.S.C. 1716), which directed the secretaries of Agriculture and the Interior to provide more uniform rules and regulations pertaining to land appraisals and establish procedures and guidelines for the resolution of appraisal disputes, including provision for use of arbitration where appropriate.

Land exchange between nonfederal landowners and the federal agencies is an alternative to acquiring nonfederal lands with appropriated funds but sometimes is difficult to accomplish, as described below.

## Effects on USFS and BLM Timber Sales

A fundamental requirement of the laws and regulations under which the federal agencies are authorized to exchange lands is that lands being acquired must be of equal value with the lands being removed from federal ownership; differences in value in any single exchange can be equalized with a cash payment in an amount that does not exceed 25% of the total transaction value. In exchanges of timberland between private parties and the federal agencies, it is not uncommon for the federal agency to seek to acquire lands that have high recreational or nontimber values from private entities in exchange for more productive commercial forest lands that meet the objectives of the private entity.

In this circumstance, the federal agency often acquires more total acres than it relinquishes, but because it conveys average or better quality timberland, usually containing some merchantable timber, to the nonfederal party, the net effect on a given USFS or BLM district can be a long-term reduction in annual timber harvest capability. In this event, companies that depend upon a predictable level of timber sales from a specific national forest or BLM district often object to such an exchange, and if the magnitude of the reduction in future timber sales is significant, such opposition can prevent the exchange from taking place.

## Effect on Local Government Revenues

A common requirement in any land exchange with a federal agency is the need to structure the parcels of lands being exchanged to avoid significant reduction in county or school district revenues resulting from federal timber sales or grazing fees. Because the distribution of federal receipts to local governments is based upon the acreage of federal ownership, a county or school district where a reduction in federal land ownership occurs might be subject to a permanent reduction in federal revenues, which might be offset in the short term by an increase in property taxes paid by the new private landowner. In cases of a significant change in historic revenues in counties or school districts where such federal revenues are an important part of the overall funding needed for schools and roads, opposition by county commissioners or school district officials might prevent an otherwise beneficial exchange from occurring. Consequently, any sizable land exchange must be structured to balance the change in federal and nonfederal ownership within county, and often school district, boundaries.

## Exchanges Involving More Than One State

The difficulty inherent in accomplishing land exchanges that cross state lines was recognized by the Public Land Law Review Commission (PLLRC) in 1970, which nevertheless recommended that "all federally owned lands otherwise available for disposal should be subject to exchange, regardless of agency jurisdiction and geographic limitation." The PLLRC encouraged the agencies to emphasize improved planning, public information practices, and public participation to gain support for large-scale exchange programs that cross state lines, such as was accomplished when public lands in Nevada were exchanged to acquire lands in California for the Point Reyes National Seashore (PLLRC, 1970).

Congress must approve any exchange of federal land that crosses state lines; therefore, rarely does any single land exchange with the federal government involve more than one state. This is particularly true if one state (and its counties) loses a significant amount of private land from the tax base and another state gains a similar opportunity. In FY 1991, however, Congress authorized a land exchange involving two states under which USFS acquired the surface ownership of 12,274 acres of inholdings in the Black Hills National Forest in South Dakota from the Homestake Mining Company in exchange for 868 acres of national forest land in Colorado (Coffin, 1990).

## Time Required to Accomplish Complex Exchanges

Experience in recent years indicates that the time and expense required to accomplish land exchanges with federal agencies has increased to the point where large-scale land exchanges involving more than a few hundred acres on each side are discouraged. To a large degree, this is the result of relatively recent laws that require federal agencies to address numerous aspects of change in federal land ownership previously not considered. Such changes include the requirements of the National Environmental Policy Act, which must be addressed in every exchange of federal land and laws requiring consultation among federal agencies regarding wildlife, archeological and historic resources, and the like. At a minimum, more time is required by an agency to complete additional requirements, including the requirement for a public notice and comment period, and sometimes, public hearings. It is not unusual for private landowners to become discouraged by the time, expense, and complexity of satisfying all of the federal requirements to accomplish an otherwise mutually beneficial exchange.

## Limits Imposed by National Forest Boundaries

With some specific exceptions, the laws that govern national forest land exchanges generally require that nonfederal lands being acquired by USFS be located within the statutory boundary of a national forest. Because only Congress can change a national forest boundary, any land exchange that involves nonfederal land outside of an existing national forest boundary requires case-by-case approval from Congress.

Exchanges that cross national forest or region boundaries usually are not encouraged by USFS because of the perception that one forest or region would be losing land or resources for the benefit of another forest or region. Similarly, unless land adjustment is actively encouraged at the higher administrative levels of the agency, local personnel might be reluctant to identify candidate exchange tracts if their district is giving up land or resources for the benefit of another USFS administrative unit.

### Conveyance of Mineral Rights

In any transaction involving the acquisition, sale, or exchange of fee title in land, the conveyance of the underlying mineral estate must be determined. The surface interest of a parcel often is owned by one landowner while the mineral estate is owned by another entity as a result of an earlier transaction.

When owned by the federal government, the mineral estate is the responsibility of BLM. Therefore, when federal lands and minerals are disposed of to a nonfederal landowner, BLM must concur in the disposition of the mineral rights, even when the surface acreage is national forest land. Although exchange of full-fee interest, including mineral rights, normally is desired by both parties to an exchange, that often is not feasible in situations where there is a large disparity in mineral values or potential for disparity between the two ownerships or when the mineral estate is owned by a third party. In such cases, partial or full mineral rights can be reserved by the federal government and the nonfederal landowner to accomplish a balanced, value-for-value exchange of surface interests.

Resolution of the differences in the mineral estate can add significantly to the complexity and time requirements associated with surface exchanges in these situations. For example, the Washington, D.C., office of USFS encourages a policy of not separating the mineral estate from the surface ownership, thereby requiring that the mineral estate be acquired from the private party in most exchanges. That policy has discouraged some private landowners from initiating otherwise desirable exchanges with USFS.

### Other Limitations on the Land Exchange Process

**Regulatory Delay**

The Federal Land Exchange Facilitation Act (43 U.S.C.A. § 1716) was enacted by Congress in 1988, but final regulations remain pending. Draft regulations were published in the *Federal Register* on August 18, 1989, and again on October 2, 1991 (*Federal Register,* Vol. 56, No. 191, 1991). In addition, no regulations and procedures are available that facilitate and encourage three-party exchanges among nonfederal landowners and more than one federal agency. For example, a nonfederal party might be willing to convey land to USFS and acquire lands of equal value from BLM. Such three-party exchanges are common among private landowners, but it is very difficult to accomplish creative three-party exchanges when more than one federal agency is involved. An exception to accomplish a similar end result for a specific situation was authorized by Congress in the 1980 Burton-Santini Act (94 Stat. 3381), where revenue from the sale of BLM lands in Clark County, Nevada, was transferred to the Land and Water Conservation Fund to help fund USFS land acquisition in the Lake Tahoe basin (BLM, 1991).

**Inadequate Identification of Lands
Unsuitable to Agency Missions
and Available for Exchange**

An examination of several land and resource management plans developed by USFS pursuant to the National Forest Management Act of 1976 (P.L. 94-588, 90 Stat. 2949) shows substantial variation in the degree of detail described to support individual national forest land adjustment plans. Notable exceptions are the plans for the Mark Twain National Forest in Missouri and the Mt. Baker-Snoqualmie National Forest in Washington, for which detailed information is presented to inform the public about land exchange opportunities (USFS, 1986, 1990a). Similarly, new resource management plans being prepared by the BLM districts in Oregon, pursuant to FLPMA, show some variation in the level of detail provided to identify land exchange opportunities at the district level, as does the Judith Valley Phillips plan (see Figure 3-1). The Roseburg District plan provides a particularly detailed explanation of that district's land exchange opportunities and clearly identifies the lands that are available to exchange as well as the lands that BLM proposes to acquire (BLM, 1991).

## Inadequate Integration With Regulatory Approaches

Exchanges or partial exchanges could serve to mitigate or supplement regulatory approaches. For example, nonfederal lands designated by USFWS as critical habitat for threatened or endangered species under ESA are not subject to acquisition from willing landowners through exchange for other federal lands, including selected national forest, public domain, or other lands administered by BLM, on a value-for-value basis that reflects the value of the nonfederal land in its current use, regardless of the critical habitat designation. The equal value or comparable value standard for land exchanges poses questions of whether the value of the land should be judged before, during, or after the imposition of regulatory restrictions. The decision to compensate for regulatory "wipeouts" of this sort is obviously a policy question that should be left to Congress.

## LAND ACQUISITION STRATEGIES AND TRANSACTIONS

### Federal Agencies: Reliance on Willing Sellers

Land most often is purchased from a willing seller by an agency. Public agencies must pay the fair market value of purchased land, although they can accept full or partial donations of land. The OMB criteria sets the existence of a willing seller as a minimum standard, with condemnation to be used only in rare instances. Nonetheless, many statutes presume an unwilling seller.

The strategy of land-acquisition campaigns and their goals might figure in the increased reliance upon willing sellers. In the Lake Tahoe experience, for example, Fink (1991) reported that "any land acquisition program must avoid increasing tensions in an already highly charged atmosphere if it is to be accepted by the local public. Often, the public strongly resists the inclusion of eminent domain in land use plans. Following the protracted disputes concerning other methods of environmental protection at Lake Tahoe, the regular exercise of condemnation to acquire land in the basin would likely have been perceived by many in the region as yet another unilateral exercise of raw governmental power." Furthermore, land conservation or preservation campaigns often proceed at a more leisurely pace than land to be acquired for other purposes, for example, to clear a highway corridor. More time and less money can reinforce the tendency to go slowly and speak softly, dealing first with willing sellers.

Moreover, land acquisition endeavors for certain conservation purposes, such as habitat or environmental protection, might not be as dependent upon complete removal of all incompatible uses, the way a highway condemnation undertaking might be. The Lake Tahoe experience is illustrative once again, because the water-quality goals were achieved by selective and partial acquisitions from willing sellers (Fink, 1991). To the degree that parcels are fungible, seeking willing sellers is good sense.

The committee recognizes that the distinction between willing and unwilling sellers can be a fine one. There are many examples of voluntary sales of property that were subtle cave-ins to the entreaties of officials. And the committee also found examples also of condemnations that worked to the advantage of the property holder. Those ranged from friendly condemnations that resolved questions of ownership or title (and deferred capital gains taxes), to arbitrated condemnations that settled frequent differences over price, to condemnations that made fair-market value available to buyers who were defrauded at the time of the original transaction.

Formal condemnations are used sparingly by the NPS, BLM, USFS, and USFWS. NPS uses condemnation principally to resolve differences over valuation; a small percentage of the agency's condemnations (perhaps 10%) is used to clear title on smaller properties. USFWS will use condemnation to settle valuation differences and ownership questions; it has used condemnation as an emergency technique to interdict threats of irreparable damage to the resources (e.g., the imminent cutting of trees) the unit was established to protect. The USFS resorts to condemnation primarily to secure access to existing federal lands. BLM uses condemnation to prevent imminent development in conservation areas. And one BLM condemnation case was necessary to exchange land with Arizona, because the state constitution does not allow exchanges with the U.S. government.

The criterion of the willing seller might be one that should be addressed on a project-by-project or function-by-function basis. Congress has made the call both ways, and the committee could identify no reason to deem the many goals of acquisition as always or never able to be accomplished by a criterion of willing seller.

## Nonprofit Organizations: Entrepreneurs and Facilitators

Partnerships between federal agencies and nonprofit organizations have grown out of perceived needs to overcome bottlenecks in the land-acquisition process and achieve flexibility unavailable to governmental entities. The amount of land acquired by the federal government through the participation of nonprofit organizations is small compared with the total land purchased; however, the parcels of land acquired often are critical. Nonprofit organizations have a greater range of options available to them than federal agencies do, and consequently, they can structure projects of broad scope and use an assortment of acquisition mechanisms not available to the federal government. The flexibility, risk-taking ability, and expertise of nonprofit organizations complement the resources of federal agencies and allow the two to accomplish together what neither can do alone.

A nonprofit organization familiar with agency procedures and regulations can facilitate the acquisition process even when the transaction involves a willing seller and an agency willing to purchase. If a nonprofit organization is involved, the title might be transferred directly to the government, or the nonprofit organization might hold the property until funding is available. A nonprofit organization also can assign an option to purchase directly to an agency.

Although federal agencies must pay the fair market value of land, a nonprofit organization can purchase property at auction, solicit full or partial donations of property, or otherwise obtain the property at a bargain rate. Nonprofit organizations also might have an advantage in dealing with landowners who are suspicious of agency appraisals, are tired of dealing with bureaucracies, or desire to use an intermediary in dealing with an agency. In addition, a nonprofit organization can offer the use of its personnel to handle the technical aspects of a transfer and thus save brokers' commissions or reduce outside brokers' commissions.

Agencies can accept full or partial donations of land, but landowners historically have not been motivated to donate property. A federal agent can inform a landowner that a bargain sale is a charitable donation, but a nonprofit organization can help structure a transaction and analyze alternatives so that a landowner obtains the maximum deduction possible while meeting financial needs.

Nonprofit organizations also can hold land for sale to an agency in increments over several years as funds become available; federal agencies cannot enter into installment purchases unless the funding has been appropriated first. Agencies can purchase options to buy property for a maximum of 1 year and $1, but nonprofit organizations can purchase options at a price that will interest a seller. At the same time, nonprofit organizations can be assembling a group of parcels or locating a willing agency to purchase the land. That is a particularly useful role: Large tracts can present opportunities to maintain an area of biological diversity or an ecosystem, but the amount of land available and deserving of protection might not fit the criteria, management objectives, and budget of a single agency.

Three conspicuous features of contemporary land-acquisition transactions are their complexity, the management challenges they present, and the necessity for cooperation with local residents. Transactions involving nonprofit organizations are complex in many ways—they take account of multiple parties (federal, state, local, and private), numerous sources of funding (private contributions, LWCF monies, local bonds, and tax revenues), extended negotiations (measured in years and even decades), and myriad legal forms and transactions (management agreements, exchanges, full- and less-than-fee purchases, trades, grants, and others).

The management difficulties associated with complex land assemblages are evident. Numerous owners have legal rights and entitlements intertwined in various ways. Decisions of one owner are linked to the others, and the choices of one manager can set a course that might be influential for all. In a variety of formal and informal ways, multiple owners are brought together in a web of common understandings, mutual expectations, and legal duty. Transactions inevitably come equipped with demands land managers must handle—more co-owners, more constraints on operations, more considerations (including extraterritorial ones) to account for, and more approaches to weigh.

Transactions negotiated by nonprofit organizations often must accommodate the residents and users already on land designated for protected status. Degrees of conflict differ, and inholders tolerated under one use regime might not be acceptable under another. But many land conservation schemes depend on the initial generosity and continued cooperation of resident landowners who assume some burdens under cooperative agreements or partial-fee dispositions.

## Hope Valley

The Hope Valley project in the Sierra Nevada Mountains of California illustrates the value of involvement by a nonprofit organization. The project encouraged multiple agency cooperation in the acquisition and management of large land holdings encompassing a variety of resources.

USFS wanted to acquire portions of land in the Hope Valley area to improve recreational opportunities within the Toiyabe National Forest, improve habitat for several endangered species, and facilitate management. The California State Department of Fish and Game was anxious to manage the riparian corridors to re-establish fish populations, and the California State Park System wanted to consolidate and improve management of existing state parks in the same area.

Portions of the land were owned by six major and several lesser landowners. Through a long series of negotiations and option purchases, the Trust for Public Land (TPL) coordinated the purchase of the land desired. In addition, TPL helped negotiate a memorandum of understanding between the various federal, state, and local agencies involved to ensure that their management protocols were compatible.

The Hope Valley undertaking was complicated by the reluctance of Alpine County to lose land from its property tax rolls. To secure the county's support, TPL purchased and donated to the county for development a parcel of land that permitted access to county services, donated funds for investment to offset a portion of the lost tax base, and arranged for USFS to exchange some publicly held lands in areas where services were available for more remote properties in private hands.

## Primerica

An acquisition can be split between agency and private use. TPL used this technique in acquiring and disposing of 80,000 acres of holdings in Wisconsin. As is often the case, the owner, Primerica (formerly American Can Company), would sell only its entire holding. The acreage was in numerous parcels scattered over eight counties. Over time, a critical 10,000-acre parcel of old-growth forest was conveyed to four groups: USFS, Wisconsin, The Nature Conservancy (TNC) for an addition to one of its preserves, and a local Indian tribe. The remaining 70,000 acres were resold to small private woodlot holders.

**Carrizo Plain**

Efforts to establish an 180,000-acre "macro-preserve" as a wildlife sanctuary in the Carrizo Plain of California's Central Valley brought together an alliance of conservation groups, oil companies, ranchers, and a variety of government agencies (Itolina, 1989). The Carrizo Plain offers prime habitat for most of the endangered species in the San Joaquin Valley (the southern portion of the Central Valley), including sandhill cranes, the San Joaquin kit fox, the blunt-nosed leopard lizard, the San Joaquin antelope squirrel, and the giant kangaroo rat. Monies to support the project came from congressional appropriations, "mitigation" fees assessed in connection with oil development on neighboring BLM lands, funds of the California Wildlife Conservation Board, and TNC, which put $2 million of its own funds to purchase 82,000 cares of ranchland owned by Oppenheimer Industries, an absentee landlord based in Kansas City, which secured the core area of the preserve (Itolina, 1989).

**Coachella Valley**

Yet another example of the role of a nonprofit organization was the recent saving of the habitat of the fringe-toed lizard in the Coachella Valley near Palm Springs, California. The California chapter of TNC wanted to establish a preserve of more than 12,000 acres for the lizard. The region was under tremendous development pressure, and TNC had to act quickly, but it had only $2 million of the $25 million required to purchase the necessary land (Martin, 1986). The desired land was owned by several people.

The initial purchase alone required $12 million and consisted of a single large parcel that the owner desired to sell only as a unit. That parcel was resold over 4 years (S. McCormick, pers. comm., CNC, 1991). TNC put together a coalition involving BLM, USFWS, and the California State Department of Fish and Game to purchase that land and adjacent parcels.

Congress already had allocated USFWS funds for that agency's portion of acquisition, because the lizard was a threatened species. BLM exchanged surplus federal land for needed land that TNC was holding. TNC then resold the surplus to replace the funds spent on the desirable land (S. McCormick, pers. comm., CNC, 1991). TNC then convinced local developers and communities to assess a development mitigation fee on each acre developed as a source of ongoing funds to support the preserve (Martin, 1986).

None of the critical projects described above would have occurred without the participation of nonprofit organizations. The Hope Valley project, which involved assembling many different parcels, would not have been possible without the ability to purchase options on parcels in the hope that the remainder could be acquired. It also required a neutral party to mediate the management agreement between the various state and federal agencies involved.

The Primerica example reflects the opposite problem. The agencies only wanted 10,000 acres; the landowner would sell only 80,000. Even if the funds were available, it would have been administratively difficult for a federal agency to purchase 80,000 acres and then sell 70,000.

The Coachella Valley example presents five complicating factors, any one of which would have barred its completion if not for participation by a nonprofit: a large parcel requiring immediate purchase, an installment sale, involvement of multiple federal and state agencies, the assembly of a preserve from multiple parcels, and private sector involvement.

## CONCLUSION

In many aspects, the work of nonprofit organizations has been the most significant development in recent federal land acquisition practices. Those groups do extraordinary and useful work as entrepreneurs, innovators, and deal makers. They act as intermediaries between sellers and government buyers. They exercise quick discovery and response capabilities that government agencies may lack. They have the skills, experience, and reputations to forge multiparcel assemblages and arrange complex transactions that cross agency boundaries and overcome public and private property distinctions. The relationship between such organizations and agencies should be structured to continue to take full advantage of the demonstrated ability of the nonprofit organizations to act swiftly to carry out priority acquisitions, while ensuring that federal acquisition priorities effectively guide the overall process.

Critics of the nonprofit organizations claim that the lands bought on behalf of the government reflect private priorities rather than public priorities, and that the transactions are lobbied through Congress. It is also said that acquisition intermediaries can buy lands at a discount and thus earn undeserved profits when they sell the land at fair market value to public entities. Some critics object to the ability of nonprofit organizations to avoid procedural constraints that apply to public agencies that are involved in land acquisition (Brookes, 1991).

Neither Congress nor the agencies are under obligation to buy specific parcels from nonprofit organizations. New issues and imaginative approaches not promoted by the land-management agencies frequently are raised in Congress through the efforts of nonprofit intermediaries, which often assume significant financial risks in helping the federal agencies with land transactions.

Nonprofit organizations dedicated to the preservation of land and other natural resources cannot fill the gap created by the lack of a comprehensive national priority system. They can and do, however, greatly enhance the land-acquisition capability of the federal agencies in a variety of ways, without cost to the taxpayer and often to the benefit of numerous federal and state agencies.

Many transactions facilitated by nonprofit organizations illustrate complexity, management coordination, and cooperation challenges involved in land acquisition efforts. Management must be coordinated among the various entities at an ecosystemic and regionwide level if any individual agency is to protect biological diversity of the resources under its jurisdiction. Some conservation schemes depend on cooperation of resident landowners. Objectives of conservation often are compatible with activities of resident populations, and local residents can benefit from protection of the renewable resource base. Nonprofit organizations might have a role in monitoring conservation easements and management agreements as well as providing training in land acquisition techniques. Finally, the partnerships between nonprofit organizations and government entities are important for protecting areas large enough to meet the requirements of wildlife species.

# 8

# The Office of Management and Budget

The OMB land acquisition priority procedure (LAPP) (Appendix B) is a mechanism for combining the requests of the agencies in the Department of the Interior and the Department of Agriculture into a single list for funding that is submitted to the Congress. LAPP places general constraints on all acquisitions; for example, the property must be within the boundaries of an existing federal conservation or recreation unit "if such boundaries are set by statute," which is customarily the case for National Park Service (NPS) and Forest Service (USFS) acquisitions. Other constraints are that the property targeted for acquisition "presents no known health/safety/liability problems (e.g., hazardous waste contamination, unsafe structures," and that there is "no indication of opposition from current owner(s)" to the proposed acquisition, although the criteria state that "condemnations may be necessary in rare instances."

For properties that meet those minimum criteria, LAPP contains a point system for comparing candidate properties. The highest number of points (80) are awarded if the proposed acquisition "provides multiple recreation opportunities . . . and is within a county with a population of one million or more"; or if "the principal benefit to be derived from the acquisition is its wetlands characteristics as defined in the Emergency Wetlands Act of 1986." Fifty points are awarded if the proposed acquisition will interdict an "*imminent* (within 2-3 years) property development that is determined by the regional or State director to be incompatible with the affected unit's authorized purpose(s)." Twenty-five points are awarded if the proposed acquisition will foreclose a "*short-to-medi-*

*um* term (within 4-8 years) property development" that is judged to be incompatible with the purpose of the unit. Points also are awarded if the acquisition satisfies other measures—preservation of the habitat of an endangered species, 40 points; preservation of a nationally significant "natural or cultural feature of a type not now represented in any Federal conservation/recreation unit," 40 points; inclusion of infrastructure that would make the property accessible to the general public and to elderly and handicapped citizens, 20 points; use of less-than-fee acquisition techniques, 10 points; improvement of manageability and efficiency of a unit, 20 points; and others.

LAPP has some mechanisms to relax the rigidity of the point system and incorporate considerations of agency judgment and preference. One provision, for example, allows an agency's assistant secretary to award points to the 20 highest priority projects that further the agency's mission.

The Land Acquisition Working Group reviews and modifies the tentative ranking of land acquisition proposals to consider, among other things, proposed exceptions to the minimum criteria and "subjective factors not taken into account in the scoring process." Examples of subjective factors are "the role of a given acquisition in a coordinated Federal/State/local effort to preserve recreation lands; the possible effect of an acquisition on State, local, or private efforts to offer competing recreation opportunities; the prospect that a private conservation group may desire to purchase the property."

## ADEQUACY OF THE LAPP CRITERIA

The LAPP criteria are subject to criticism on several grounds. The primary difficulties stem from the ambition of the aim and the choice of the means. The aim is to facilitate comparisons of the acquisition proposals of numerous federal land managers. The means to achieve this is comparison by a numerical ranking system.

One of the most obvious inadequacies of the LAPP criteria is that they compare acquisition alternatives without regard to the specific purpose of the acquisition. Although a numerical ranking scheme obviously facilitates comparisons after the problem of assigning numbers is overcome, valuation judgments always will be at issue.

## The Minimum Criteria

Even the minimum criteria show how difficult it is to develop a single set of standards for all federal land acquisitions. For example, the criterion that restricts acquisitions to within or contiguous to existing units diminishes the options of NPS and USFS but not of the Fish and Wildlife Service (USFWS), which theoretically can look for land nationwide.[1] On the other hand, recent study of land acquisition in the Lake Tahoe area shows that the simple legislative technique of limiting acquisition programs to defined geographic areas can give the endeavor a precision and focus that otherwise might be lacking (Fink, 1991).

Whether the federal government should purchase property contaminated with hazardous waste is a difficult decision. But that criterion excludes consideration of ecosystem restoration and mitigation of hazards on otherwise desirable properties. Indeed, land of this sort is now for sale at Love Canal. The criterion that specifies a willing seller is contrary to the condemnation of private property for public purposes sanctioned in scores of statutes applicable to the public lands.

The LAPP criteria acknowledge the connection between acquisition and use with a threshold requirement that the cost of infrastructure necessary to make the property accessible, safe, and usable by the general public does not exceed 10% of the estimated purchase price. Costs of management are pertinent to acquisition decisions, although some sites might merit building a substantial infrastructure, just as some sites might justify mitigating a hazard.

---

[1] Acquisitions nationwide are themselves constrained. If USFWS were instructed by Congress to "maximize the preservation of biological diversity" with its acquisition funds, the agency conceivably might be tempted to make strategic purchases of land in Latin America or elsewhere outside of North America. The acquisition policies that established a network of military bases around the globe obviously were not confined by considerations of domestic geography. Whether a similar principle will govern the establishment of worldwide ecological bases is a political question.

## The Twelve Ranking Criteria

### Prevents Imminent Development

This criterion highlights the opportunistic aspects of land acquisition. Imminence of development usually is a manifestation of several features (e.g., location, value, size, and control of access) that influence the value of a particular parcel. The development decision thus might have important multiplier effects. The Washington Department of Wildlife's recent purchase of a cattle ranch in the Methow Valley, which a developer had proposed transforming into a resort, for example, is likely to influence future development in profound ways (*Seattle Times*, 1991).

A strategy of opportunism gives priority to near-term problems and diminishes the role of long-term goals and plans. Purchases to head off imminent development, moreover, tend to be high in local friction and costs, which can take a toll on acquisition budgets.

### Provides Multiple Recreation Opportunities Close to Population Centers

This criterion is limited and overly simple. It does not purport to measure the quality of the recreational experience. It provides only an indication of the possible demand for recreation without considering the extent to which this demand is already being met by federal or other public recreation areas. The important question pertains to the available supply of recreational opportunities relative to population. As written, the criterion gives priority for land acquisition to places like King County, Washington—which includes Seattle and has a wealth of readily available opportunities in national forests and state lands—at the expense of, for example, St. Louis, which is two counties away from the Mark Twain National Forest and has no other nearby extensive public forests.

### Preserves Endangered and Threatened Species Habitat

This criterion assigns 40 points if the acquisition would preserve the habitat of an endangered species and 30 points if it would preserve the habitat of a threatened species. But the aggregate number of points that can be assigned to a property because of other criteria can exceed 300. Thus, the endangered species criterion shows particularly well the effect of disparate acquisition programs serving distinct needs: regardless of importance, this criterion could not be the primary motivation for an acquisition. This is a curious outcome given the strong congressional commitment to the protection of species in the Endangered Species Act.

Properties that might be regarded as the highest priorities from the point of view of endangered species benefits can rank well below other properties whose value to endangered species is much less but whose scores are enhanced by extraneous considerations. Thus, if a priority list of proposed acquisitions for the federal endangered species program were constructed, it is likely that such a list would be substantially rearranged when it was passed through the filter of the LAPP criteria.

Moreover, beyond introducing a systematic bias against acquisitions for certain protected species (e.g., those located far away from major urban areas), the criteria also might reward factors inimical to the conservation of certain species. For example, some species, like the piping plover and the Chesapeake tiger beetle, depend upon undisturbed beach habitats. The conservation of those species is likely to depend upon fairly strict controls on a variety of recreational activities; yet, under the uniform criteria, more recreational opportunities translate into higher overall scores. Another example is the desert tortoise, whose conservation likely depends upon restricting off-road vehicle use. The existing criteria either systematically hinder the acquisition of land for those species, or encourage the tolerance of incompatible uses as part of the price of acquisition.

This criterion also suffers from a failure to adapt to rapid change. In recent times, an increasing amount of nonfederal land is being proposed as critical habitat for threatened and endangered species—millions of acres alone for the spotted owl. When the Land and Water Conservation Fund (LWCF) Act was enacted in 1964, neither Congress nor the federal agencies could foresee the extent to which the protection of threatened and endangered species habitats would become a major land management objective of the 1990s. Future federal acquisition priorities must recognize this, and the acquisition criteria and the congressional appropriations process must be able to respond to this important conservation need.

Insofar as the value of a proposed acquisition for endangered species is concerned, this criterion is a crude determination. A property either is habitat for a threatened or endangered species or it isn't; all endangered species habitat properties receive 40 points, and all threatened species habit properties receive 30 points. Yet, there are other considerations that could be extremely important in ranking proposed endangered species acquisitions. For example, an acquisition that helps avert a conflict over a proposed development might be viewed as a higher priority than one that does not. An acquisition that encompasses several endangered species is presumably a higher priority than one that encompasses only one. An acquisition within designated critical habitat is likely more important than one outside critical habitat. An acquisition of a larger tract or of a critical corridor should be accorded more weight. Those and other considerations could be used to fine-tune acquisition priorities for the endangered species program, but they are too detailed to receive any attention in the broad-brush criteria of the LAPP system.

**Preserves a Nationally Significant Natural or Cultural Feature of a Type Not Now Represented in Any Federal Unit**

This criterion presumes some compilation or list of features necessary to complete a preservation system. Although work of this sort has been done by NPS, nothing approaching a consensus (scientific or otherwise) has been reached on the components of such a preservation system. At the same time, recognition of this criterion shows the tension between stability and dynamics that attends any attempt to restate acquisition criteria. Knowledge and values about what is worth preserving change rapidly, and acquisition practice must be responsive to that.

This criterion reflects also the view that preservation of nationally significant natural and cultural features is a justification for federal acquisition. The committee generally shares this view, but federal acquisitions are only part of a complex web of state, federal, local, and private property holdings. Important features of the type adverted to in this criterion might be able to be protected by other means.

## Preserves Biologically Valuable Wetlands or Riparian Habitat

An acquisition can be awarded 80 points for this criterion, double that of protecting endangered species. Like many of the criteria, the term wetlands has been given a definition, although considerable discretion is involved in distinguishing among different types of wetlands.

As with the endangered species criterion, the wetlands criterion shows how difficult it is to reconcile an acquisition program with a regulatory program. Under Section 404 of the Clean Water Act, for example, wetlands are protected by law and use restrictions that can be imposed on private owners without compensation and within a broad range of conditions. Indeed, the higher the public values of wetlands, the lower the likelihood that payment is required as a formal legal matter under the takings clause. Among the policy questions presented are whether acquisitions should be deferred until regulation is proven inadequate; whether acquisition and regulatory programs focus on different kinds of properties; whether acquisitions achieve goals that regulation cannot; and whether a high priority (or high point) acquisition program undercuts the effectiveness of the regulatory program by raising expectations of a buyout.

## Includes Infrastructure for Access for the Public and Handicapped People

The assignment of points for improved access is quite plausible. Acquiring key land parcels and rights-of-way that can provide public access to large areas of otherwise inaccessible federal land is extremely important. For example, at the request of the Bureau of Land Management (BLM) and USFS, the Keystone Center conducted a policy dialogue in 1988 on the problems associated with obtaining public access to federal land where such access was blocked by adjacent private ownership (Keystone Center, 1989). That report identifies the confusion, frustration, and confrontation that often attends barriers to access to the public lands. It recommends incorporating access needs in the planning processes of government entities, as well as developing the innovative and practical means to improve access.

Infrastructure presumably is defined elsewhere, but it might include disparate items, such as highways on the one hand and curb-cuts for wheelchairs on the other.

## Expands Units With Recent Rapid Growth in Visitor Use

This criterion effectively makes increased visitor use a goal, even when it might be something to be avoided. It suggests the need for setting visitor use goals that vary considerably from one federal area to another. The criterion also should reflect the relationship between federal and other public use areas in serving recreation users.

## Improves Manageability and Efficiency of a Unit

Although pertinent to acquisition practice, this criterion suffers from the same bluntness that permeates the other criteria. The points available under this criterion do not provide for differentiation between large projects with sizable administrative savings and small projects that provide few, if any, savings. The ability to identify administrative cost savings is particularly important when considering land exchanges, where both parties might be able to reduce the long-term costs of property-line surveys, road development, and other land management expenses.

## Results in Federal Savings in Acquisition Costs Through the Use of Land Exchanges, Donations, or Other Alternatives

Alternatives to acquisition must be considered. Particularly for large projects and during an era of constrained federal budgets, the ability to accomplish high-priority acquisitions without using appropriated funds should be encouraged, so that LWCF monies can be used most effectively.

Land exchanges could help reduce the perception in many western states that the percentage of federal land ownership is too high and increasing, to the long-term detriment of the local property-tax base and economy. Federal acquisition of land for conservation purposes in those areas might be accomplished more easily if citizens and governments were aware that other federal lands were being made available to the private sector or to a state agency.

Nonetheless, it is very difficult to manipulate a point system to ensure that a high-priority, noncash acquisition opportunity ranks above an identical opportunity that requires the use of LWCF appropriations. An acquisition of that sort is not necessarily superior; exchanges and donations are quite complex. Furthermore, the criterion is based on percent savings in acquisition cost rather than a more relevant absolute dollar savings.

## Involves Federal Acquisition of Less Than Full-Fee Title to the Property

Federal acquisition practice has not fully exploited the opportunities for less-than-fee acquisition. But this important proposition is buried in an evaluation system that gives a slight mark-up (10 points) to acquisitions that use a less-than-fee approach. The criterion also does not account for the quality of the lands acquired. Insistence upon less-than-fee analyses would be better placed in the procedures of the individual agencies rather than in the document that sets acquisition priorities.

## Involves Significant Nonfederal Partnership

The purpose of this criterion is not clear. If the purpose is to obtain financial support from others, points should be awarded in relation to the proportion of costs borne by the partners. The current allocation of 5 points for each partner effectively means that points are awarded for adding complexity to the arrangement.

**Ranks 20 Highest Priority Projects
According to Agency Mission**

Although the criteria give weight to individual agency priority rankings, agency priorities with special criteria that differ from the priorities of the administration often are reversed. For example, in fiscal year 1991, the highest ranking USFWS project under the LAPP criteria was ranked 32nd on the agency priority list.

## REFLECTION OF AGENCY
## MISSIONS AND AUTHORITIES

For purposes of illustration, the LAPP criteria are discussed below as they relate to the missions and land acquisition authorities of a single agency—USFS. USFS has aquisition authority under LWCF, the National Trails Act, the Wild and Scenic Rivers Act, and the Federal Land Policy and Management Act. Purposes of land acquisition under those authorities include land that provides access to national forests over nonfederal lands. Other acts establishing specific units, such as national recreation areas, might have additional land acquisition authority.

Parts of the national forests serve specific purposes, and Congress has provided specific land acquisition authority in addition to the general authority of the Weeks Law and the broad outdoor recreation acquisition authority of the LWCF Act. The criteria for setting acquisition priorities, thus, must serve an array of purposes and uses, some of which deserve greater priority in land acquisition than others, despite the even treatment apparently assigned by the Multiple-Use and Sustained-Yield Act.

For purposes of this analysis, the committee divided the LAPP criteria into four groups:

- *Protection criteria* (e.g., prevents imminent development, preserves endangered and threatened species, preserves natural or cultural features not now in any federal area, preserves biologically valuable wetlands or riparian areas)
- *Recreation criteria* (provides multiple recreation opportunities close to population centers; expands units with recent rapid growth in visitor use)

- *Administrative criteria* (includes infrastructure for access by public and handicapped people; improves manageability)
- *Cost criteria* (leads to federal cost savings through land exchanges and donations; involves federal acquisition of less than full-fee title; involves significant nonfederal partnership)

## Protection and Recreation Criteria

The protection and recreation criteria are related to, but not entirely coincident with, the statutory missions for the national forests. They fail to address timber, grazing, watershed protection, and mining uses of the national forests, although this might be appropriate at the present time in view of the increased attention being assigned to wildlife and ecosystem concerns relative to other conservation purposes. The criteria also fail to address the specifics of the several statutes that designate areas of the national forests as wilderness, wild and scenic rivers, national trails, and other areas, which should have special attention in land acquisition programs.

The protection and recreation criteria also fail to address ecosystem-management concerns on the national forests, especially those related to avoiding forest fragmentation. Those are important in the context of protecting endangered species in forested areas, such as the northern spotted owl, but do not receive attention in the criteria.

The protection criteria in one group and the recreation criteria in another reflect the dichotomy in objectives that were to be met by the LWCF. That poses problems of the sort mentioned above, whereby protection of endangered species could be assigned low priority overall because one criterion does not combine well with any of the recreation-related criteria and might well be wholly at odds with them.

The recreation criteria do not fit recreation on the national forests well. Priority is given to opportunities for multiple kinds of recreation in counties with populations of 1 million or more. This could give acquisition priority to counties that already have a wealth of recreational opportunities at the expense of counties that do not.

## Administrative Criteria

These criteria do not relate to conservation goals but are concerned with how conservation goals are met. The criterion for fair access is related mainly to recreation. Assigning the same weight (20 points) to existing infrastructure to make the area accessible to the handicapped and to making the area accessible to the general public is generally inappropriate for the national forests, although it might be appropriate in some cases. Infrastructure for providing access to the national forests for the general public usually means high-cost roads; access for the handicapped might mean trails modified to allow wheelchairs, which are usually much less costly than roads. Thus, an area with good roads might be given a lower rating for acquisition because it does not have handicapped access, although access could be provided later at costs much below those for building basic road access.

The criterion that gives priority to units with rapidly growing visitor use also does not fit the national forests very well. Most of the national forests support extensive recreation. It is unlikely that such recreation will be limited by the area of land available. It is more likely to be limited by restrictions on supporting facilities, such as parking areas and campgrounds. This criterion appears more relevant to national recreation areas in urban areas (e.g., Gateway NRA) than to national forests.

Manageability could have some relevance to meeting conservation goals. It could be used to give weight to acquiring inholdings, which are common on the national forests and pose administrative and cost problems. Where such inholdings are relevant to meeting conservation goals, they presumably would be assigned priority under the protection criteria. The day-to-day problems of maintaining land lines, controlling use, and providing access suggest that this criterion deserves attention. It should not, however, be thought of in the same way as those criteria that are related to conservation goals.

## Cost Criteria

These criteria are aimed at getting the largest area or greatest value of land for whatever amount is spent. They account for a total of 50 of the maximum of 420 points that could be awarded, exclusive of the 150 points for use by the assistant secretary. Presumably, the expected acquisition costs against which savings are measured are based on fair market value. The committee is not so concerned with the weight assigned to cost savings (50 points), but with the applicability of the mechanisms that are considered (less than full-fee title and partnerships) relative to the conservation goals that are to be met.

Botkin (1990) suggests three categories of natural areas that should be maintained: no-action wilderness, which is needed as a baseline for science; preagricultural wilderness, in which the goal is to maintain the appearance of the landscape as first viewed by explorers; and conservation areas set aside to conserve biological diversity. The latter two categories generally would require active, sometimes intensive, management to meet the specific goals for which the areas were maintained.

The inclusion of criteria for less than full-fee title acquisitions and partnership in the criteria suggest substantially more important considerations than costs that should be addressed.

## Conclusions

As applied to the national forests, the criteria are imprecise and overly simple. They do not address the real tensions between recreation and protection as the two major goals of the LWCF or the conflicts between imperatives, such as endangered species protection, priorities for completing congressionally designated areas, systems (e.g., wilderness), and other less-absolute objectives. They mix criteria for protection and recreation goals with administrative and cost criteria. And they miss the great variety of conditions in the national forests: solid blocks versus fragmented parcels, spectacular versus ordinary landscapes, isolated versus well-traveled lands, and economically useful lands versus rock and icefields.

The scheme ranks potential areas in and around the national forests with points assigned for the various goals and other factors without any precision. And none may be necessary. The agencies' claims that Congress accepts their rankings for at least 70% of the acquisitions might reflect the agencies' knowledge of what will and what will not be acceptable in Congress more than it does the efficacy of the ranking system.

# 9

# Conclusions and Recommendations

The Committee on Scientific and Technical Criteria for Federal Acquisition of Lands for Conservation was asked to (1) review the criteria and procedures by which the four major federal land management agencies acquire public lands for conservation purposes; (2) assess the historic, public policy, and scientific bases for the agencies' land-acquisition criteria; (3) compare them with those of nongovernmental land conservation groups; (4) assess the effectiveness of the federal land-acquisition programs; and (5) evaluate the extent to which the agencies have objective methods for ranking potential acquisitions.

The four land management agencies—the National Park Service (NPS), the Fish and Wildlife Service (USFWS), the Bureau of Land Management (BLM), and the Forest Service (USFS)—have wide-ranging missions and mandates. Land acquisition is a tool available to each of them but is not the major mission of any of them. Except for BLM, whose authority for acquiring lands is relatively new, each agency has its own ranking system for land acquisition. Superimposed on, and developed from, the agencies' rating systems is an interagency system for setting acquisition priorities used by the Office of Management and Budget (OMB) for preparing the land-acquisition portion of the president's annual budget.

The committee found that the various approaches used by each agency to rate potential land acquisitions generally are based on systematic criteria that reflect the agencies' basic missions. The missions themselves, however, are complex, reflecting sometimes conflicting or confusing goals and a long history of federal public land policies. The committee compared the agencies' criteria with those used by nongovernmental land-conservation groups and found that the private groups typically have straightforward land-acquisition goals and programs. The committee believes that the agencies can benefit from review and understanding of the nongovernmental groups' programs, but the nongovernmental criteria are not directly transferable to the federal land-acquisition programs.

Criteria for federal land-acquisition programs must change from time to time to keep pace with evolving agency missions. For example, the rationale for the Land and Water Conservation Fund (LWCF), the main source of appropriations for federal land acquisitions, has changed since it was established in 1964. Its focus has evolved over time to include acquiring land for various conservation purposes. Even the meaning of conservation has evolved over the years, with changes in scientific understanding and increasing pressures on the nation's land base.

With these changes have come revised understanding of the meaning of acquisition and of the forces that affect its usefulness as a conservation tool. The separation of wholly private and wholly public lands has become less distinct with increasing reliance since the 1960s on regulations and less-than-fee acquisitions to accomplish public objectives. The committee believes that such reliance will increase in the future, because land is limited, and meeting a range of public and private goals simultaneously is becoming increasingly difficult.

Understanding of conservation needs is expanding as scientific knowledge grows. Conservation programs are coming to grips with the recognition of the importance of maintaining biological diversity, the potential for changes in global climate, the value of working landscape approaches to regional protection, and other environmental variables. Sustaining natural conditions over some significant landscapes is one approach to preserving biological diversity. In view of existing landscape patterns and land uses, a mix of land-acquisition techniques, such as conservation easements and less-than-fee purchases, will need to be used creatively and cooperatively across agency lines and with nonfederal governments and private landowners.

The committee believes that the conservation objectives of the federal government can best be met by continuing the commitment of the land-management agencies and OMB to a rational planning approach for setting land-acquisition priorities. That means defining current federal lands in relation to conservation needs, determining what additional lands or ownership in land would contribute to these needs, and establishing priorities so that land-acquisition appropriations are allocated efficiently. At the same time, the existing process can be improved. The recommendations that follow are intended to improve the current system for setting land-acquisition priorities and the use of various means of acquiring ownership in conservation lands.

## GOALS

### Structuring OMB and Agency Criteria

*OMB, NPS, BLM, USFS, and USFS should separate the current national ranking system for funding acquisition priorities into at least three categories: outdoor recreation resources, natural resources protection, and cultural heritage protection—the three major purposes of federal land acquisition. Other categories might be needed, especially where Congress has designated portions of the federal lands to protect specific kinds of resources, such as wilderness areas, wild and scenic rivers, and historic and archeological sites.*

Each agency should develop individual criteria to rank its own acquisitions, because no single set of criteria will work to satisfy fully the different agency missions. The OMB method for setting federal acquisition priorities among the priorities of individual agencies forces nonadditive criteria into a single composite ranking. This skews results in favor of potential land acquisitions that meet some of each purpose and against acquisitions that would best meet specific purposes. The current approach also emphasizes certain considerations at the expense of others and diminishes the agencies' ability to fulfill their legislative charges.

Outdoor recreation and wildlife protection on the same tract of land can be incompatible. Yet, the OMB criteria award points for each in a single ranking system for federal land acquisitions. For example, public recreation is assigned as many as 80 points in setting priorities, while protecting endangered and threatened species is assigned a maximum of

40 points. This favors acquisition of areas that can expect high recreation use as well as have endangered or threatened species. The present ranking system applies the same criteria nationwide and across all agencies, despite variance in regional needs and agency missions. A system that ranks the Civil War Battlefield at Gettysburg against the mission blue butterfly does not have the flexibility to serve all the purposes encompassed by the agencies responsible for land acquisition.

Some latitude is possible with the current composite criteria if the assistant secretaries use their discretionary points (150 points for the highest priority tract) to shift the balance in favor of an acquisition when this is needed to ensure that a specific goal is met. But as long as a single set of criteria is used to meet all acquisition purposes, the system is flawed. The federal agencies, with their varied missions, need more than a single set of criteria.

One approach would be parallel ranking systems for each major purpose, leaving Congress to decide on the total amount of appropriations to allocate to each purpose. Alternatively, Congress could establish dedicated funds for each major federal acquisition purpose. If the agencies ranked their requests by major purpose, secondary purposes could be recognized and given some weight, but each acquisition would be counted toward the main purpose. The report of the President's Commission on Americans Outdoors makes note of similar dedicated funds: the Wallop-Breaux amendments to aid sport fishing and recreation on federal lands, the Reclamation Act to support irrigation projects, the Highway Trust Fund, and the Historic Preservation Fund. Although this approach might establish a rationale for a ranking scheme, the committee cautions that bases for dedicated funds quickly become outdated.

The agencies' missions in relation to land acquisition need to be made explicit to help clarify the various criteria. Broad categories of shared agency goals, such as those used in the OMB criteria, hide some important distinctions. All four agencies provide outdoor recreation opportunities, but some of those provided by USFS and BLM are not provided by NPS or the USFWS. For example, all-terrain vehicles and motorized trail bikes, as well as hunting, are permitted on large parts of the national forests and BLM lands, but usually not in the national parks. The same sort of distinction applies to wetlands—should USFWS acquisition of prairie potholes to support migratory waterfowl be afforded the same priority as BLM acquisition in a wild and scenic river corridor to meet ecosystem management objectives?

Recreational and biological conservation goals can be separated while recognizing that conservation of biotic resources fosters outdoor recreation opportunities. But the two goals are incompatible in some cases and complementary, or at least nonconflicting, in others. The current OMB criteria imply that they always are complementary.

## Acquisition Strategies and Techniques

*Agencies should use the widest possible range of land-protection strategies in formulating acquisition proposals, from public ownership to land-use regulation, alternatives to fee-simple land acquisition, exchanges, public-private and interagency arrangements, partnerships, cross-boundary planning, and other techniques.*

The federal land base is not used to the fullest extent possible in meeting goals for which land acquisition is a tool because of a lack of interagency planning, multiple missions and mandates, and agency behavior that assigns low priority to some missions.

NPS, BLM, USFS, and USFWS focus on the lands for which they are directly responsible and historically have emphasized some parts of their missions at the expense of others. As a result, opportunities for meeting broad recreation and resource conservation goals with the current mix of federal, state, and private lands often are overlooked, and expansion of the federal land base is seen as the only solution. For example, meeting landscape-level habitat needs of wide-ranging wildlife species requires attention from all four of the federal agencies. It also requires consideration of arrangements other than fee-simple land acquisition to connect habitat on lands of different agencies.

Over time, strengthening the incentives for partnerships and other public-private options as complements to fee-simple acquisition will require direction from agency leaders and support from Congress. Another technique worth agency experimentation is reservation of conservation interests and reversionary policies in property dispositions, such as was done with the reservation of mineral rights in the late nineteenth and early twentieth centuries. Conditions of this sort appear in property dispositions ranging from the railroad land grants to the Rykers Act that authorized the flooding of the Hetch-Hetchy Valley in Yosemite. Obviously, property not disposed of need not be repurchased.

The majority of federal land acquisitions today are in-fee purchases. However, like interest retention, less-than-fee acquisition techniques sometimes allow limited dollars to be stretched to fulfill acquisition goals. The committee is aware that use of these techniques by the agencies has not always been successful and that practical limitations are associated with less-than-fee acquisitions. Although the committee does not believe that in-fee acquisition should be abandoned in favor of less-than-fee alternatives in all cases, it also does not believe that historical limitations of less-than-fee techniques should be accepted as an inescapable policy constraint. Vaguely worded less-than-fee agreements could be rewritten with specificity, and agreements too short in duration could be extended into the future. Unenforceable agreements could be made enforceable—by including monetary penalties, using loss-of-property reversions for violations, or using third-party enforcement techniques. Most contemporary federal environmental laws invite third-party enforcement through the mechanism of citizen suits. A similar technique might prove useful in the context of conservation easements. Nongovernmental organizations often are important in arranging public-private cooperation. And nonprofit organizations might be able to negotiate contractual arrangements for monitoring and training.

Congress and the executive branch should consider measures to remove some of the barriers to the more widespread use of land exchange. Those measures might include, for example

- Completion by USFS and BLM of final regulations for implementation of the Federal Land Exchange Facilitation Act (43 U.S.C. 1716);
- Development of regulations and procedures to facilitate and encourage three-party exchanges among nonfederal landowners and more than one federal agency;
- Improvement in the ability of federal agencies to accomplish exchanges that cross state lines;
- Strengthening of the training and development of land-exchange specialists within the federal agencies and assignment of the most experienced individuals to the agencies' top-priority land-exchange projects;
- Examination of ways to supplement local government or school district revenues on a one-time basis when there is a change in federal landownership;
- Recommendations from USFS and BLM for ways in which the

cumbersome and time-consuming exchange process could be streamlined.

## Landscape and Ecosystem Protection

*Land-acquisition criteria should be expanded to include landscape pattern analysis, which typically includes land-use and land-cover data and measures certain factors, such as patch characteristics, vegetation types, ecological trends, and hydrologic and socioeconomic interactions with the resources. Land uses in an entire watershed should be considered in the design of reserves.*

Traditional acquisition practice evaluates individual parcels without considering regional attributes, including biogeographical and landscape patterns. A complementary approach is needed among the four agencies, as is cooperation to protect ecosystems and habitats that transcend agency jurisdictional boundaries.

Comparative evaluations of parcels are distorted if they miss the regional contexts and the ecological dynamics to which the properties are subject. Usually, acquisitions that provide corridors, connections, and linkages between similar landscapes and habitats are enhanced in biological value. In the same way that a strategically situated piece of property can provide access to public lands for human users, habitats in proper configurations can facilitate the persistence, movement, and dispersal of native biota.

In addition, individual tracts of protected land can be affected by external factors; for example, agricultural runoff from lands outside the refuge resulted in poisoned waterfowl on the Kesterson Wildlife Refuge. And if global climate change does occur, it might accelerate migration of certain species. Future public land acquisitions must be sensitive to the dynamics of landscape patterns and uses.

One approach for interagency cooperation might be to develop criteria for land uses in an entire watershed. Such criteria would identify the sustainability of regional land uses and identify tradeoffs between meeting current needs and maintaining options for the future. An important consideration for land acquisition is the extent to which a particular acquisition contributes to maintaining those options. Consideration of land uses throughout a watershed is critical for management of any aquatic resources, such as wild and scenic rivers and endangered fish stocks.

### Representative Natural Areas

*NPS, BLM, USFS, and USFWS should prepare an overall strategic plan that identifies land-acquisition needs for establishing and protecting representative natural areas on federal lands that can provide scientific baselines for judging the effects of human actions on the natural environment.*

The federal land-acquisition process does not adequately address the need for protecting natural areas as scientifically credible baselines to measure the effects human use has on resources. Of the four land-management agencies, USFS has been the leader in developing a scientifically credible system; it has established a system of more than 250 natural research areas in the national forests. That system is about 60% complete. A full system of research areas should include ecosystems that are best represented by federally owned areas on lands managed by the other three agencies, as well as some areas not in federal ownership.

Standards for determining what areas are in the system and how they should be protected should be consistent among agencies. An interagency committee, parallel to one in the late 1970s, could help to establish consistent guidelines and provide a useful mechanism for agreeing on areas to be included as research natural areas.

The system should represent ecosystems as completely as possible. Land acquisition needed for such a system of research areas probably is modest relative to other programs.

## PROCEDURES

### Planning and Acquisition

*The agencies should develop and use long-term land-acquisition plans that can be used to identify priorities and opportunities for interagency cooperative efforts.*

The agencies should take into account regional conservation needs as well as social effects of acquisitions on local landowners and communities and provide a mechanism for public participation. The multiyear perspective of such plans would enable Congress to judge how well the agencies fulfill their missions and facilitate effective evaluation of the cumulative effects of land acquisition.

Overall vision and long-term planning that consider cumulative as well as social effects are needed for the acquisition of federal land. That would allow a variety of public-policy and scientific objectives to be considered (e.g., economic benefits and costs to communities, provision of corridors, protection of watersheds, and consideration of other available recreational opportunities in the area). Social impact analysis (SIA) is an essential tool in successful land-use planning and should be conducted to identify problems in advance and compare alternatives, provide a mechanism for public participation, record public needs and site-specific interests, provide baseline measurements for future comparison, identify poorly understood cumulative effects, review spatially and temporally remote interests, and identify possibilities for mitigation.

Federal acquisitions can have large effects, positive and negative, on local residents and communities. Some of the negative effects are lost tax revenues, disrupted traditional community patterns, and dislocated human activities. Positive effects include increased revenues from tourism, more recreational opportunities, increased adjacent property values, and protection of the renewable resource base. Although the creation of a national park and protection of natural resources should not be contingent on creating or preserving employment, SIA can be used to identify the distribution of costs and benefits to minimize the costs and provide the public with clear and reliable information regarding the implications of public land policy. A comprehensive checklist of social effects could be developed to determine the social significance of individual acquisitions.

The conservation objectives of many acquisitions often are compatible with a variety of activities of the resident population. In many cases, co-management between federal authorities and the resident population is a realistic goal; the local residents often have the most to gain by extending and protecting the renewable resource base. SIA also can reduce costs by identifying problems and comparing alternatives in advance.

The OMB criteria and the agency submittals provide an annual agenda for acquisition, but they appear to start from a set of criteria that are only partly related to agency missions. In the effort to get common criteria across agency lines, much of the sense of individual agency missions is lost. Furthermore, a multiyear perspective seems to be lacking, as are signs of a systematic overview of acquisition needs other than at the field level.

The OMB criteria assign no special weight to congressionally designated areas, such as wilderness areas, wild and scenic rivers, and national trails. But agency land-use plans often emphasize congressionally designated areas. Agency land-use plans vary widely; some have criteria for land acquisition that appear nowhere in the national criteria, which implies that information from the field is not used at the national level.

### Improving Information for Decision Making

*The federal land-acquisition program for conservation should have a solid information base as part of a systematic approach to achieving its goals.*

That information base should enable the land-management agencies and Congress to determine the extent to which conservation needs are being met and to identify gaps in meeting those needs.

Gap analysis entails examining the distribution of key elements of biological diversity relative to areas now under some type of protective ownership. A geographic information system (GIS) consists of the computer hardware and software for manipulating spatially distributed data. GIS is an especially powerful tool for planning and acquisition of conservation lands and can be applied in the study of environmental processes, analysis of trends, and predictions of the results of planning decisions. The methodologies of gap analysis and GIS are widely used today in resource planning decisions. They need to be applied to setting priorities for federal land acquisition.

Information should be assembled for NPS, BLM, USFS, and USFWS in a common GAS. The agencies should continue to refine and expand their applications of gap analysis and GAS. Data gathering should be improved, extended, and directed with a view toward applications in gap analysis and GIS.

Information needed to determine priorities for land acquisition such as maps of landownership, land use, critical habitats for wildlife and endangered species, natural areas, water availability, and vegetation is scattered among several federal and local agencies and is not in a form that is readily accessible to decision makers. A long-term view of acquisition hinges on an inventory of current landholdings defined in relation to major management objectives and on the identification of areas that should be acquired in fee or in part. Additional data that are incorporated with increasing frequency in GISs include social science information, such as human population change and other census data.

The agencies do have inventories of current landholdings, often as part of their land-use plans. For example, typical USFS land-use plans show allocation of national forest lands to various major land-use categories, the agency's ownership in relation to other owners, and lands to be acquired and lands to be exchanged or otherwise disposed of. But they usually lack an interagency and regional view of land-acquisition needs and an objective view of what could be accomplished with less-than-fee acquisitions and interagency land uses. Descriptions in land-use plans of what could be accomplished through acquisitions, for example, of wildlife corridors that would connect units managed by different agencies, including state agencies, would be helpful.

A four-agency information base for a conservation land-acquisition program should be drawn from existing information bases, such as the Environmental Protection Agency's Environmental Monitoring and Assessment Program, state natural history surveys, state GAS programs, and the NPS biosphere reserve program. Such an information base could be similar to one proposed by the Conservation Foundation (1985), which suggested a three-part program for setting priorities for new national parks:

- First is a register, or nationwide list of natural and cultural resources, of "sites worthy of special management." That register would not set priorities, but would identify the universe of resources worthy of consideration.
- Second, a set of thematic inventories should be undertaken, combing the register for sites relevant to a particular opportunity or concern. For example, an inventory might be done of sites important to floral biological diversity, sites threatened by climate change, or corridors connecting species populations amid fragmented landscapes.

- Third, Congress should receive periodic reports evaluating acquisition opportunities. Such a report, updated every 2 or 3 years, would include a statement of the federal backlog, a summary of recent purchases, an estimate of backlog costs, and an analysis of the land market.

Inventories identifying private lands that might be subject to full-fee or partial acquisition by the federal government are a sensitive issue. The committee notes, however, that problems posed by such inventories appear to have been addressed in agency land-use plans that identify specific areas of private lands as desirable for federal acquisition. If a systematic approach to setting priorities for federal conservation land acquisition is to be accomplished, it is clear that some private lands must be identified and information collected on their suitability for meeting specific needs. The committee believes that the experience of agencies identifying such lands in land-use plans can be used to guide collection of information necessary for useful inventories for further land acquisition without interfering with the privacy of landowners.

The major advances for identifying gaps in protected lands and the quality of public and private lands deserve to be recognized. Those new methodologies, however, are heavily dependent on the adequacy of existing data and maps for such basic questions as ownership, inventories, population trends, distribution of species, and so on. The need for information systems to support land- and resource-management decisions is not confined to the federal program. State-agency data bases are notoriously incomplete, scattered, incompatible, and inaccessible.

## Funding

*For long-term planning and consistent adherence to a set of criteria the LWCF needs adequate and predictable funding.*

Funding for the LWCF has fluctuated dramatically; for example, appropriations in 1978 were approximately $800 million; in 1982, less than $200 million; and in 1991, approximately $375 million. Appropriations to the states have been variable as well, but generally have been much less than the maximum of 60% allowed by the LWCF Act; in 1982, the states received only 2% of the total appropriations. Such variation makes planning for land acquisition difficult. National planning should be attentive to local planning. National criteria should be tied to criteria used in local land-use plans and should give weight to congressionally designated areas. Funding must take account of those factors.

### Emergency Acquisitions

*Congress should consider mechanisms, such as providing discretionary LWCF funding for dealing with emergencies and unexpected opportunities.*

Discretionary funding would allow the secretaries of the Department of the Interior and the Department of Agriculture to take advantage of unexpected opportunities or respond to unwelcome threats to resources.

During the several years normally consumed by the process of identifying lands for acquisition, planning, and study, conditions can change; e.g., prices often rise and qualities fall. In addition, sudden events (such as development threats and purchase opportunities) are common in the course of land acquisition. Parcels that were unavailable might become available because of tax delinquency, foreclosure, or death. Parcels might come under imminent threat of development when the land is sold. In cases such as those, the lack of discretionary funding can mean that an agency is unable to acquire the property.

The tension between carefully considered and opportunistic actions is a stumbling block in the development of comprehensive acquisition criteria. The rise of nonprofit organizations and the active role Congress takes is evidence of that. The objections to emergency acquisitions, such as unaccountable and ad hoc actions, can be met by requiring stringent after-the-fact explanation and accounting.

### Monitoring Acquisitions and Re-evaluating Criteria

*Acquisitions should be monitored periodically to determine if the purposes for acquisition have been realized. Criteria also should be periodically re-evaluated in light of changes in holdings, climate and biological resources, demographics, scientific knowledge, and policy and political values.*

Acquisitions need to be monitored to assess their effectiveness in achieving goals within the context of agency missions. In a sense, USFS and BLM already do this in their periodic revisions of land-use plans; land-use plans are reviewed and revised every 10 to 15 years. Criteria for acquisitions should be reviewed in light of the agencies' changing missions.

## State and Local Issues

*Precise criteria should be developed to meet national outdoor recreation and conservation needs in setting federal land-acquisition priorities. NPS, BLM, USFS, and USFWS should consider the needs and resources of state, local government, and Indian tribal lands in federal land-use plans, as well as the role of state and local governments in providing outdoor recreation, especially as these are defined in statewide comprehensive outdoor recreation plans.*

The potential role for state and local government lands in providing outdoor recreation is given too little attention in setting priorities for federal land acquisition. The conservation needs of Indian tribes have not been addressed by federal or state programs.

In passing the LWCF Act, Congress assigned an important role to state and local governments in providing opportunities for public outdoor recreation. As much as 60% of the annual appropriations can be granted to the states for land acquisition and development of recreation areas. In recent years, a much lower proportion usually has gone to the states, in part because administration budgets have not asked for more.

The committee found that NPS, BLM, USFS, and USFWS acquisition programs pay little attention to nonfederal outdoor recreation opportunities on other lands. One result is that recreation is emphasized in federal land-acquisition priorities even in areas where such opportunities might be provided on state and local lands. This skews the federal land-acquisition priorities against other reasons for acquisition, such as protection of wildlife and endangered species.

Full funding of the state grant portion of the LWCF appropriations would relieve the pressure on federal agencies to provide recreation opportunities to meet local needs. Yet more is required than a simple expansion of entitlements. The program should be merit- and goal-driven. Development of a clear set of recreation guidelines to meet national recreation needs would focus federal land acquisitions on high-priority national needs.

## Incentives for Private Landowners

*Because conservation land needs cannot be satisfied through public land-acquisition programs alone, efforts should be made to develop partnerships and other mechanisms of cooperation with private landowners to achieve goals that have been realized to date primarily through acquisition.*

Even the largest nature reserves, if left alone, probably will suffer major die-offs of species in a few hundred or a few thousand years. Size demands are greater if the reserve is located in a disturbance-prone environment or if it is intended to accommodate migrations of protected species.

By any measure, future conservation needs of the nation will outpace any efforts by federal land buyers to satisfy those needs with traditional acquisition practices. NPS, BLM, USFS, and USFWS have recorded impressive backlogs of properties that satisfy acquisition criteria but await funding; current acquisition practice makes only small dents in the formal specifications of acquisition needs. On top of this, the under-sized nature of many biological preserves and the space essential for effective wildlife conservation underscore the futility of relying on simply spending more federal dollars to create habitats of sufficient size.

The recommendation above is a natural outgrowth of the recognition that any successful campaign to protect biological diversity cannot be constrained by traditional demarcations between public and private properties. Historically, land managers respond to incentives. Less-than-fee acquisitions are one useful technique for extending habitat protection, and the committee cannot say what other forms of incentives might be useful. But a nation that has paid farmers not to grow pigs may yet find the will to pay them to grow owls, eagles, or hedgerows.

## Acquisition Intermediaries

*NPS, BLM, USFS, and USFWS should continue to take advantage of the ability of nonprofit organizations to act swiftly to secure properties until an agency can acquire them. Federal acquisition priorities should guide the process, and the transactions should be in accordance with federal guidelines that control dealing with nonprofit organizations.*

The amount of land acquired by the federal government through the participation of nonprofit organizations is small compared with the total amount of land acquired. But the nonprofit organizations do play an important role in the acquisition of critical tracts; they provide agencies with important flexibility in certain situations and can be key when timing or flexibility is essential. As early as 1970, the Public Land Law Review Commission recommended that the federal agencies use alternative acquisition techniques to combat the price escalation of lands required for federal programs.

The federal agencies have developed guidelines for transactions with nonprofit conservation organizations that emphasize the need to ensure that federal priorities guide federal acquisitions. The guidelines establish procedures governing disclosure and reimbursement to nonprofit organizations when they sell land to the government. They make clear also that the nonprofit organizations do not act as agents of the government and that the agencies and the Congress decide whether to buy specific tracts. The committee believes that the guidelines provide a useful framework for the relationships between the agencies and the land-acquisition intermediaries.

# References

Aiken, B. 1988. The Kruger: A Supreme African Wilderness. Jersey: Afropix Publishers.

Allen, T.F., and T.B. Starr. 1988. Hierarchy: Perspectives for Ecological Complexity. Chicago: University of Chicago Press.

Anderson, T.L., ed. 1983. Water Rights: Scarce Resource Allocation, Bureaucracy, and the Environment. San Francisco: Pacific Institute.

Anderson, T.L., and D.R. Leal. 1991. Free Market Environmentalism. Boulder, Colo.: Westview Press.

Anderson, J.R., E.E. Hardy, J.T. Roach, and R.E. Witmer. 1976. A land use and land cover classification system for use with remote sensor data. USGS Professional Paper 964. U.S. Geological Survey, Reston, Virginia.

Arnold, R. 1982. At the Eye of the Storm: James Watt and the Environmentalists. Chicago: Regnery Gateway.

Backus, E.H., R.M. Alfaro, L.F. Corrales, Q. Jimenez, L.H. Elizondo, and W.H. Soto. 1988. Costa Rica: Assessment of the Conservation of Biological Resources. Fundacion Neotropica and Conservation International, San Jose, Costa Rica, and Washington, D.C.

Baker, W.L. 1989a. Landscape ecology and nature reserve design in the Boundary Waters Canoe Area, Minnesota. Ecology 70:23-35.

Baker, W.L. 1989b. A review of models of landscape change. Landscape Ecol. 2:111-133.

Barlowe, R. 1965. Federal programs for the direction of land use. Iowa Law Review 50:337-367.

Barrett, T.S., and P. Livermore. 1983. The Conservation Easement in California. Covelo, Calif.: Island Press.

Bean, M.J. 1983. The Evolution of National Wildlife Law. New York: Praeger Publishers.

Beede, S.F. 1991. Le Parc National des Cévennes. Pp. 100-106 in Resident People and National Parks: Social Dilemmas and Strategies in International Conservation, P.C. West and S.R. Brechin, eds. Tucson: University of Arizona Press.

Blahna, D.J. 1986. Social Bases for Resource Conflicts in Areas of Reverse Migration. Paper presented at the First National Symposium on Social Science in Resource Management, Oregon State University, Corvallis, Oregon. May 12-16.

BLM (Bureau of Land Management). 1990. Recreation 2000: A Strategic Plan. Bureau of Land Management, U.S. Department of the Interior. Washington, D.C.: U.S. Government Printing Office.

BLM (Bureau of Land Management). Undated. Fish and Wildlife 2000: A Plan for the Future. Washington, D.C.: U.S. Department of the Interior.

BLM (Bureau of Land Management). 1991. Judith Valley Phillips Resource Management Plan and Environmental Impact Draft Statement, July, 1991. Montana State Office, Bureau of Land Management, U.S. Department of the Interior.

Bormann, F.H., and G.E. Likens. 1991. Pattern and Process in a Forested Ecosystem: Disturbance, Development, and the Steady State Based on Hubbard Brook Ecosystem Study. New York: Springer Verlag.

Botkin, D.B. 1990. Discordant Harmonies: A New Ecology for the Twenty-First Century. New York: Oxford University Press.

Bremer, T. 1984. Portrait of land trusts. Pp. 17-24 in Land-Saving Action, R.L. Brenneman and S.M. Bates, eds. Covelo, Calif: Island Press.

Brookes, W. 1991. Greenlining: Back door to limiting our use of land? The Washington Times, Editorials Commentary. January 17.

Brumbach, B.C., and R.A. Brumbach. 1988. The Nexus of Land Acquisition and Regulation: Alternative Techniques for Protecting the Public Interest. Urban Law & Policy 9:319-.

Burgess, R.L., and D.M. Sharpe, eds. 1981. Forest Island Dynamics in Man-Dominated Landscapes. New York: Springer-Verlag.

# REFERENCES

Burrough, P.A. 1986. Principles of Geographical Information Systems for Land Resources Assessment. Oxford: Clarendon Press.

Castle, E.N. 1982. Land, people, and policy: A national view. J. Soil Water Conserv. 18-20.

Chambers, D. 1986. Development and application of a pilot geographic information system for the U. S. Forest Service: Tongass National Forest. Pp. 162-171 in Proceedings of the Geographic Information System Workshop, Atlanta, Ga.

Chavez, M.J. 1987. Public access to landlocked public lands. 39 Stan. L. Rev. 1373-.

Ciriacy-Wantrup, S.V. 1952. Resource Conservation: Economics and Policies. Berkeley: University of California Press.

Ciriacy-Wantrup, S.V. 1985. Economics and policies of resource conservation. Pp. 207-230 In Natural Resource Economics: Selected Papers, R.C. Bishop and S.O. Andersen. Boulder, Colo.: Westview Press.

Clark, J.S. 1988. Effect of climate change on fire regimes in northwestern Minnesota. Nature 334:233-235.

Clawson, M. 1973. Historical overview of land-use planning in the United States. Pp. 23-54 in Environment: A New Focus for Land-Use Planning, D.M. McCallister, ed. Washington, D.C.: National Science Foundation.

Clawson, M., and R.B. Held. 1957. The Federal Lands. Baltimore, Md.: The Johns Hopkins University Press.

Coffin, J.B., ed. 1990. Colorado-South Dakota exchange. Public Land News 15:24:3.

Coggins, C.G. 1991. Public Natural Resources Law. June.

Coggins, G.C., and C.F. Wilkinson. 1987. Federal Public Land and Resources Law, 2nd ed. Mineola, N.Y.: The Foundation Press, Inc.

Cohen, F.S., ed. 1982. Handbook of Federal Indian Law, 1982 Edition, R. Strickland, ed. Charlottesville, Va.: Michie Bobbs-Merrill.

Comptroller General of the United States. 1966. Report on Examination into Certain Proposed Land Exchanges for the Point Reyes National Seashore in California. U.S. General Accounting Office, Washington, D.C.

Conservation Foundation. 1985. National Parks for a New Generation: Visions, Realities, Prospects. Conservation Foundation, Washington, D.C.

Coulson, R.N., C.N. Lovelady, R.O. Flamm, S.L. Spradling, and M.C. Saunders. 1991. Intelligent geographic information systems for natural resource management. Pp. 153-172 in Quantitative Methods in Landscape Ecology, M.G. Turner and R.H. Gardner, eds. New York: Springer-Verlag.

CEQ (Council on Environmental Quality). 1990. Environmental Quality. Twenty-First Annual Report of the Council on Environmental Quality, Washington, D.C.

Crespi, M. 1984. The potential role of national parks in maintaining cultural diversity. Pp. 303-309 in National Parks, Conservation, and Development: The Role of Protected Areas in Sustaining Society, J.A. McNeely and K.R. Miller, eds. Washington, D.C.: Smithsonian Institution.

Crumpacker, D.W., S.W. Hodge, D. Friedley, and W.P. Gregg, Jr. 1988. A preliminary assessment of the status of major terrestrial and wetland ecosystems on Federal and Indian lands in the United States. Conservation Biology 2:103-115.

Cunningham, R.A. 1967. Scenic easements in the highway beautification program. 45 Denver Law Journal 167-.

Dames and Moore. 1982. Social Impacts and Effect of CAWCS Plans. Final report prepared for the Bureau of Reclamation, U.S. Department of the Interior, Washington, D.C.

Dana, S.T., and S.K. Fairfax. 1980. Forest and Range Policy, 2nd ed. San Francisco: McGraw-Hill.

Davis, M.B. 1989. Insights from paleoecology on global change. Bull. Ecol. Soc. Am. 70:222-228.

Davis, M.B., and D.B. Botkin. 1985. Sensitivity of cool-temperate forests and their fossil pollen record to rapid temperature. Quat. Res. 23:327-340.

Davis, F.W., D.M. Stoms, J.E. Estes, J. Scepan, and J.M. Scott. 1990. An informations systems approach to the preservation of biological diversity. Internat. J. Geograph. Inform. Syst. 4:55-78.

DeAngelis, D.L., and J.C. Waterhouse. 1987. Equilibrium and nonequilibrium concepts in ecological models. Ecological Monographs 57:1-22.

Delcourt, H.R., and P.A. Delcourt. 1991. Quaternary Ecology: A Paleoecological Perspective. New York: Chapman & Hall.

Delcourt, H.R., P.A. Delcourt, and T. Webb. 1983. Dynamic plant ecology: The spectrum of vegetational change in space and time. Quat. Sci. Rev. 1:153-175.

Diamond, J. 1984. "Normal" extinctions of isolated populations. Pp. 191-246 in Extinctions, M.H. Nitecki, ed. Chicago: University of Chicago Press.

Diamond, J.M. 1986. The design of a nature reserve system for Indonesian New Guinea. Pp. 485-503 in Conservation Biology: The Science of Scarcity and Diversity, M.E. Soulé, ed. Sunderland, Mass.: Sinauer Assoc., Inc.

DOI (U.S. Department of the Interior, Bureau of Indian Affairs). 1979. Increasing Participation by Indian Tribes in the LWCF. Task force report prepared by the Bureau of Indian Affairs and the Heritage Conservation and Recreation Service. August.

Dregne, H.E. 1991. Report on the Cost of Global Desertification (Temporary Title), Desertification Control PAC, United Nations Environmental Program. New York: The United Nations.

Drury, N.B. 1946. Private in-holdings in the national park system. Land Policy Review 9:3-8.

Dunn, C.P., D.M. Sharpe, G.R. Guntenspergen, F. Stearns, and Z. Yang. 1991. Methods for analyzing temporal changes in landscape pattern. Pp. 173-198 in Quantitative Methods in Landscape Ecology: The Analysis and Interpretation of Landscape Heterogeneity, M.G. Turner and R.H. Gardner, eds. New York: Springer-Verlag.

Elfring, C. 1989. Preserving land through local land trusts. BioScience 39:71-74.

Ellis, D. 1989. Environments at Risk: Case Countries of Impact Assessment. New York: Springer-Verlag.

Emanuel, W.R., H.H. Shugart, and M.P. Stevenson. 1985a. Climatic change and the broad-scale distribution of terrestrial ecosystem complexes. Clim. Change 7:29-43.

Emanuel, W.R., H.H. Shugart, and M.P. Stevenson. 1985b. Response to comment: Climatic change and the broad-scale distribution of terrestrial ecosystem complexes. Clim. Change 7:457-460.

Fink, R.J. 1991. Public land acquisition for environmental protection: Structuring a program for the Lake Tahoe basin. Ecology Law Quarterly 18:485-557.

Finsterbusch, K. 1985. State of the art of social impact assessment. Environment and Behavior 17:193-221.

Fleming, D. 1972. Roots of the new conservation movement. Perspectives in American History 6:7-91.

Forman, R.T.T., and M. Godron. 1986. Landscape Ecology. New York: John Wiley & Sons.

Foresta, R. 1984. America's National Parks and Their Keepers. Washington, D.C.: Resources for the Future.

Foresta, R.A. 1987. New national parks: Lessons from the United States and Canada. Forum for Applied Research and Public Policy 2:95-107.

Foster, D.R. 1988. Disturbance history, community organization, and vegetation dynamics of the old-growth Pisgah Forest, southwestern New Hampshire, USA. J. Ecology 6:105-134.

Frankel, O.H., and M.E. Soulé. 1981. Conservation and Evolution. Cambridge, England: Cambridge University Press.

Franklin, J.F., and R.T.T. Forman. 1987. Creating landscape patterns by forest cutting: Ecological consequences and principles. Landscape Ecology 1:5-18.

Fox, S. 1986. The American Conservation Movement: John Muir and His Legacy. Madison: The University of Wisconsin Press.

Frederick, D.O. 1991. Nature in the hands of accountants. National Wetlands Newsletter 13(4):4-7.

Freudenburg, W.R. 1986. Social impact assessment. Annu. Rev. Sociol. 12:451-478.

Freudenburg, W.R., and R. Gramling. 1992. Community impacts of technological change: Toward a longitudinal perspective. Social Forces 70:937-955.

GAO (General Accounting Office). 1979. The Federal Drive to Acquire Private Lands Should be Reassessed. CED-80-14. Report by the Comptroller General of the United States, U.S. General Accounting Office, Washington, D.C. December 14.

GAO (General Accounting Office). 1981. Federal Land Acquisition and Management Practices. Report to Senator Ted Stevens by the U.S. General Accounting Office. CED-81-135. Washington, D.C.: U.S. Government Printing Office.

GAO (General Accounting Office). 1988. Endangered Species: Management Improvements Could Enhance Recovery Program. Report to the Chairman, Subcommittee on Fisheries and Wildlife Conservation and the Environment, Committee on Merchant Marine and Fisheries,

House of Representatives. GAO/RCED-89-5. Washington, D.C.: U.S. Government Printing Office. December.

GAO (General Accounting Office). 1989. National Wildlife Refuges: Continuing Problems with Incompatibile Uses Call for Bold Action. GAO/RCED-89-196. Washington, D.C: U.S. Government Printing Office. September.

Gardner, R.H., B.T. Milne, M.G. Turner, and R.V. O'Neill. 1987. Neutral models for the analysis of broad-scale landscape pattern. Landscape Ecol. 1:19-28.

Geisler, C.C. 1983. The New Lay of the Land. Introduction to Who Owns Appalachia? Lexington, Ky.: University of Kentucky Press.

Geisler, C.C. 1992. Adapting Social Impact Assessment to Protected Area Development. Paper prepared for the Global Environmental Facility of the World Bank. Washington, D.C. January.

Gilligan, J.P. 1953. The Development of Policy and Administration of Forest Service Primitive and Wilderness Areas in the Western United States. Ph.D. Dissertation. University of Michigan, Ann Arbor.

Glicksman, R.L., and G.C. Coggins. 1984. Federal recreational land policy: The rise and decline of the Land and Water Conservation Fund. 9 Columbia J. Environ. Law 125-236.

Golley, F.B. 1984. Managing parks and reserves as ecosystems: A report from a Workshop organized by the International Association for Ecology (INTECOL) and sponsored by the U.S. National Park Service. Institute of Ecology, University of Georgia, Athens.

Gordon, H.S. 1954. The economic theory of a common-property resource: The fishery. J. Polit. Econ. 62:124-142.

Gottlieb, A.M., ed. 1989. The Wise Use Agenda: The Citizen's Policy Guide to Environmental Resource Issues. Bellevue, Wash.: The Free Enterprise Press.

Greeley, W.B. 1972. Forests and Men. Salem, N.H.: Ayer Co. Pubs., Inc.

Greer, S. 1984. Rationalization, Power and the Forest Service: A Case Study of Conflict over Mount Rogers National Recreation Area, Virginia. Unpublished Ph.D. dissertation. University of Kentucky, Lexington, Kentucky.

Grumbine, R.E. 1990. Viable populations, reserve size, and federal lands management: A critique. Conservation Biology 4:127-134.

GSA (General Services Administration). 1989. Summary Report of

Real Property Owned by the United States Throughout the World as of September 30, 1989. Office of Government Wide Real Property Relations, U.S. General Services Administration, Washington, D.C.

Gustafson, G.C. 1983. Who Owns the Land? A State and Regional Summary of Landownership in the United States. ERS Staff Report AGES830405. Natural Resources Economics Division, Economic Research Service, U.S. Department of Agriculture, Washington, D.C. April.

Hamilton, N.R. 1985. Legal authority for federal acquisition of conservation easements to provide agricultural credit relief. 35 n3 Drake Law Review 477-527.

Harmon, D. 1991. National park residency in developed countries: The example of Great Britain. Pp. 33-39 in Resident Peoples and National Parks: Social Dilemmas and Strategies in International Conservation, P.C. West and S.R. Brechin, eds. Tucson: University of Arizona Press.

Harris, L.D. 1984. The Fragmented Forest: Island Biogeography Theory and the Preservation of Biotic Diversity. Chicago: University of Chicago Press.

Harris, L.D., and R.D. Wallace. 1984. Breeding bird species in Florida forest fragments. Proceedings of the Annual Conference of Southeastern Associations Fish and Wildlife Agencies 38:87-96.

Harris, L.D. 1988. Landscape linkages: The dispersal corridor approach to wildlife conservation. Transactions of the North American Wildlife and Natural Resources Conference 53:595-607.

Harris, L.D., and J. Scheck. 1991. From implications to applications: The dispersal corridor principle applied to the conservation of biological diversity. Pp. 189-220 in Nature Conservation 2: The Role of Corridors, D.A. Saunders and R.J. Hobbs, eds. New York: Surrey Beatty & Sons Pty. Ltd.

Hays, S.P. 1959. Conservation and the Gospel of Efficiency: The Progressive Conservation Movement, 1890-1920. Cambridge, Mass: Harvard University Press.

Hays, S.P. 1987. Beauty, Health and Permanence: Environmental Politics in the United States, 1955-1985. Cambridge: Cambridge University Press.

Hemmet, S.A. 1986. Parks, people, and private property: The National Park Service and Environmental Domain. Environ. Law 935:

957-958.

Heritage, J. 1974. On private holdings in national forests. Christian Science Monitor (Sept. 24):15.

Hiss, T. 1990. The Experience of Place. New York: Alfred A. Knopf, Inc.

Hodgson, M.E., J.R. Jensen, H.E. Mackey, Jr., and M.C. Coulter. 1987. Remote sensing of wetland habitat: A woodstork example. Photogrammetric Engineering and Remote Sensing 54:1601-1607.

Hoffman, S.M. 1989. Open space procurement under Colorado's Scenic Easement Law. 60 U. Colorado Law Review 383.

Holling, C.S., ed. 1978. Adaptive Environmental Assessment and Management. International Institute on Applied Systems Analysis (IIASA). New York: John Wiley & Sons.

Holling, C.S. 1986. The resilience of terrestrial ecosystems: Local surprise and global change. Pp. 292-317 in Sustainable Development of the Biosphere, W.C. Clark and R.E. Munn, eds. Cambridge: Cambridge University Press.

Holling, C.S. 1992. Sustainability: The Cross-Scale Dimension. Unpublished manuscript, Department of Zoology, University of Florida, Gainesville, Florida.

Hoose, P.M. 1981. Building an Ark: Tools for the Preservation of Natural Diversity Through Land Protection. Covelo, Calif.: Island Press.

Houck, O.A. 1988. Ending the war: A strategy to save America's coastal zone. 47 Maryland Law Review 358-405.

Hough, J.L. 1991. Social impact assessment: Its role in protecting area planning and management. Pp. 274-283 in Resident Peoples and National Parks: Social Dilemmas and Strategies in International Conservation, P.C. West and S.R. Brechin, eds. Tucson: University of Arizona Press.

Howell, B.J. 1984. Accommodating local interests in national park planning: A case study from the Big South Fork National River and Recreation Area (Kentucky-Tennessee). In Proceedings of the First World Conference on Cultural Parks, Mesa Verde, Colorado.

Huntley, B.J. 1988. Conserving and monitoring biotic diversity: Some African examples. Pp. 248-260 in Biodiversity, E.O. Wilson, ed. Washington, D.C.: National Academy Press.

Huntley, B.J., E. Ezcurra, R. Fuentes, K. Fujii, P. Grubb, W. Haber,

J. Harger, M. Holland, S. Levin, J. Lubchenco, H. Mooney, V. Neronov, I. Noble, H. Pulliam, P. Ramakrishnan, P. Risser, O. Sala, J. Sarukhan, and W. Sombroek. 1991. A Sustainable Biosphere: The Global Imperative. INTECOL Number 20: Special Issue. Aiken, S.C.: International Association for Ecology.

Itolina, D. 1989. Plain dealing. Sierra Jan-Feb:143.

Iverson, L.R. 1988. Land-use changes in Illinois, USA: The influence of landscape attributes on current and historic land use. Landscape Ecol. 2:45-61.

Jackson, D.A., and H.H. Harvey. 1989. Biogeographic associations in fish assemblages: Local vs. regional processes. Ecology 70:1472-1484.

Johnson, L.B. 1990. Analyzing spatial and temporal phenomena using geographical information systems: A review of ecological applications. Landscape Ecol. 4:31-44.

Johnston, K.M. 1987. Natural resource modeling in the geographic information system environment. Photogrammetric Engineering and Remote Sensing 53:1411-1415.

Johnston, C.A., N.E. Detenbeck, and G.J. Neimi. 1988. Geographic information systems for cumulative impact assessment. Photogrammetric Engineering and Remote Sensing 54:1609-1615.

Keystone Center. 1989. Policy Dialogue on Access to Federal Lands. Final Report. The Keystone Center, Keystone, Colo. March 7.

Keystone Center. 1991. Final Consensus Report of the Keystone Policy Dialogue on Biological Diversity on Federal Lands. Keystone Center, Keystone, Colo. April.

Klockenbrink, M. 1991. The new range war: The desert as enemy. New York Times (August 20):A-15.

Knight, D.H. 1987. Parasites, lightning, and the vegetation mosaic in wilderness landscapes. Pp. 59-83 in Landscape Heterogeneity and Disturbance, M.G. Turner, ed. New York: Springer-Verlag.

Knowlton, C. 1986. Cultural impacts of New Mexico and West Texas reclamation projects. Southwestern Review 5:13.

Krummel, J.R., R.H. Gardner, G. Sugihara, R.V. O'Neill, and P.R. Coleman. 1987. Landscape patterns in a disturbed environment. Oikos 48:321-324.

Land Rights Letter. 1991. Land Rights Letter: For Americans dedicated to preserving our heritage of private property rights. Land Rights

Letter 1(July):1-8.
Land Trust Alliance. 1991-1992. Survey results in 1991-1992 National Directory of Conservation Land Trusts. Land Trust Alliance, Washington, D.C.
Lefkovitch, L.P., and L. Fahrig. 1985. Spatial characteristics of habitat patches and population survival. Ecological Modelling 30:297-308.
Leopold, A.S. 1925. Wilderness as a form of land use. J. Land Publ. Util. Econ. 1:398-404.
Leopold, A.S. 1933. Game Management. New York: Charles Scribner's Sons.
Lewis, D.G. 1978. Who Owns the Land? A Preliminary Report for the Southern States. Staff Report NRED/ERS 80-10. Natural Resource Economics Division, U.S. Department of Agriculture. Washington, D.C.: U.S. Government Printing Office.
Lewis, J.A. 1980. Landownership in the United States, 1978. AIB-435. Division of Natural Resource Economics, U.S. Department of Agriculture. Washington, D.C.: U.S. Government Printing Office. April.
Llewellyn, L.G. 1974. The social impact of urban highways. Pp. 89-108 in Social Impact Assessment. Man-Environment Interactions: Evaluations and Applications, C.P. Wolf, ed. Knoxville: University of Tennessee Press.
Lovejoy, T.E., R.O. Bierregaard, Jr., A.B. Rylands, J.R. Malcolm, C.E. Quintela, L.H. Harper, K.S. Brown, Jr., A.H. Powell, G.V.N. Powell, H.O.R. Schubart, and M.B. Hays. 1986. Edge and other effects of isolation on Amazon forest fragments. Pp. 257-285 in Conservation Biology: The Science of Scarcity and Diversity, M. Soulé, ed. Sunderland, Mass.: Sinauer Associates.
Lubchenco, J., A.M. Olson, L.B. Brubaker, S.R. Carpenter, M.M. Holland, S.P. Hubbell, S.A. Levin, J.A. MacMahon, P.A. Matson, J.M. Melillo, H.A. Mooney, C.H. Peterson, H.R. Pulliam, L.A. Real, P.J. Regal, and P.G. Risser. 1991. The sustainable biosphere initiative: An ecological research agenda. Ecology 72:371-412.
MacArthur, R.H. and E.O. Wilson. 1967. The Theory of Island Biogeography. Princeton, N.J.: Princeton University Press.
Madden, J.L. 1983. Tax incentives for land conservation: The charitable contribution deduction for gifts of conservation easements. 11 nl

Boston College Environmental Affairs Law Review 105-148.

Madson, C. 1988. The Land and Water Conservation Fund: Paying for a Nation's Playgrounds. The Nature Conservancy Magazine:4-7.

Mandelker, D.R. 1982. Land Use Law. Charlottesville, Va.: Michie Co.

Mansfield, M. 1991. On the cusp of property rights: Lessons from public land law. Ecology Law Q. 18:43-104.

Marshall, R. 1930. The problem of the wilderness. Scientific Monthly 30:141-148.

Martin, G. 1986. The Preservation Corporation. Reprinted from The Robb Report, The Magazine for Connoisseurs, Volume 10(6).

Matuszeski, W. 1966. Less Than Fee Legal Devices for Open Space Preservation in Metropolitan Areas: Feasibility and Implementation. April. Unpublished.

May, R.M. 1973. Stability and Complexity in Model Ecosystems. Princeton, N.J.: Princeton University Press.

McIntosh, R.P. 1985. The Background of Ecology, Concept and Theory. Cambridge: Cambridge University Press.

Milne, B.T. 1988. Measuring the fractal dimension of landscapes. Appl. Math. and Comput. 27:67-79.

Milne, B.T. 1991. Lessons from applying fractal models to landscape patterns. Pp. 199-235 in Quantitative Methods in Landscape Ecology: The Analysis and Interpretation of Landscape Hererogeneity, M.G. Turner and R.H. Gardner, eds. New York: Springer-Verlag.

Milne, B.T., K.M. Johnston, and R.T.T. Forman. 1989. Scale-dependent proximity of wildlife habitat in a spatially-neutral Bayesian model. Landscape Ecol. 2:101-110.

Montana Land Reliance/Land Trust Exchanges. 1982. Private Options: Tools and Concepts for Land Conservation. Covelo, Calif.: Island Press.

Montgomery, C., and G.M. Brown, Jr. 1991. Economics of Species Preservation: A Marginal Analysis of Northern Spotted Owl Viability. Paper presented at annual meeting of the Western Economics Association, Seattle, Wash.

Mooney, H.A., and M. Godron, eds. 1983. Disturbance and Ecosystems. New York: Springer-Verlag.

Morine, D.E. 1990. Good Dirt: Confessions of a Conservationist. Chester, Ct.: The Globe Pequot Press.

Morris, S.W. 1982. How are land-saving decisions made? Establishing criteria, keeping priorities. Pp. 189-191 in Private Options: Tools and Concepts for Land Conservation, Montana Land Reliance and Land Trust Exchange, eds. Covelo, Calif.: Island Press.

Murphree, M.W. 1991. Research on the institutional contexts of wildlife utilization in communal areas of eastern and southern Africa. Pp. 137-145 in Wildlife Research for Sustainable Development, J. Grootehuin, S.A. Njugunn, and P.W. Kat, eds. Kenya Agricultural Research Institute, Kenya Wildlife Service, National Museums of Kenya, Nairobi, Kenya.

Negri, S. 1989. New BLM preserve: The San Pedro Riparian area. Arizona Highways 65(4):18-33.

Neilson, R.P., and L.H. Wullstein. 1983. Biogeography of two southwest American oaks in relation to atmospheric dynamics. J. Biogeogr. 10:275-297.

Netherton, R.D. 1979. Environmental conservation and historic preservation through recorded land-use agreements. 14 Real Prop. Prob. & Tr. J. 540-.

Newmark, W.D. 1986. Species-area relationship and its determinants for mammals in western North American national parks. Biol. J. Linn. Soc. 28:83-98.

Newmark, W.D. 1987. A land-bridge island perspective on mammalian extinctions in western North American parks. Nature 325:430-432.

Noss, R.F. 1983. A regional landscape approach to maintain diversity. BioScience 33:700-706.

Noss, R.F. 1987. Corridors in real landscapes: A reply to Simberloff and Cox. Conserv. Biol. 1:159-164.

Noss, R.F. 1990. Why we need to think big: Biodiversity and conservation in Greater Yellowstone. Greater Yellowstone Report 7(Fall):1,4-6.

Noss, R.F., and L.D. Harris. 1986. Nodes, networks, and MUMs: Preserving diversity at all scales. Environ. Manage. 10:299-309.

NPS (National Park Service). 1988. Management Policies. National Park Service, U.S. Department of the Interior, Washington, D.C. December.

NPS (National Park Service). 1990. Keepers of the Treasures: Protecting Historic Properties and Cultural Traditions on Indian Lands.

A Report on Tribal Preservation Funding Needs submitted to Congress. Interagency Resources Division, Branch of Preservation Planning, National Park Service. Washington, D.C.: U.S. Department of the Interior.

NRC (National Research Council). 1992a. Science and the National Parks. National Academy Press: Washington, D.C.

NRC (National Research Council). 1992b. Restoration of Aquatic Ecosystems. Science, Technology, and Public Policy. National Academy Press: Washington, D.C.

Nyland, R.D., W.C. Zipperer, and D.B. Hill. 1986. The development of forest islands in exurban central New York state. Landscape and Urban Planning 13:111-123.

Odum, E.P., and M.G. Turner. 1990. The Georgia landscape: A changing resource. Pp. 137-164 in Changing Landscapes: An Ecological Perspective, I.S. Zonneveld and R.T.T. Forman, eds. New York: Springer-Verlag.

O'Leary, J.T. 1976. Land use redefinition and the rural community: Disruption of community leisure space. J. Leisure Research 8:263-274.

O'Neill, R.V., D.L. DeAngelis, J.B. Waide, and T.F.H. Allen. 1986. A Hierarchical Concept of Ecosystems. Princeton, N.J.: Princeton University Press.

Olwig, K.R. 1980. National parks, tourism, and local development: A West Indian case. Human Organization 39:22-31.

Osborne, L.L., and M.J. Wiley. 1988. Empirical relationships between land use/cover and stream water quality in an agricultural watershed. J. Environ. Mgmt. 26:9-27.

Ostrom, E. 1990. Governing the Commons: The Evolution of Institutions for Collective Action. New York: Cambridge University Press.

Palmeirim, J.M. 1988. Automatic mapping of avian species habitat using satellite imagery. Oikos 52:59-68.

Pastor, J., and W.M. Post. 1988. Response of northern forests to $CO_2$-induced climate change. Nature 344:55-58.

PCAO (President's Commission on Americans Outdoors). 1986. Report and Recommendations to the President of the United States. The President's Commission on Americans Outdoors. Washington, D.C.: U.S. Government Printing Office. December.

PCAO (Presidents's Commission on Americans Outdoors). 1988. Report and Recommendations to the President of the United States. The President's Commission on Americans Outdoors. Washington, D.C.: U.S. Government Printing Office. December.

Perdue, C., Jr., and N.J. Martin-Perdue. 1979-80. Appalachian fables and facts: A case study of the Shenandoah National Park removals. Appalachian J. 7:84-104.

Perdue, C., Jr., and N.J. Martin-Perdue. 1991. To build a wall around these mountains: The displaced people of Shenandoah. The Magazine of Albemarle County History 49:48-71.

Peters, R.L., and J.D.S. Darling. 1985. The greenhouse effect and nature reserves. BioScience 35:707-717.

Peters, R.L., and T.E. Lovejoy, eds. 1992. Global Warming and Biological Diversity. New Haven, Conn.: Yale University Press.

Pickett, S.T.A., and J.N. Thompson. 1978. Patch dynamics and the design of nature reserves. Biol. Conserv. 13:27-37.

Pickett, S.T.A., and P.S. White, eds. 1985. The Ecology of Natural Disturbance and Patch Dynamics. Orlando, Fla.: Academic Press.

PLLRC (U.S. Public Land Law Review Commission). 1970. One Third of the Nation's Land. A Report to the President and the Congress by the Public Land Law Review Commission. Washington, D.C.: U.S. Government Printing Office. June.

Pulliam, H.R. 1988. Sources, sinks, and population regulation. Am. Naturalist 132:652-661.

Powell, J.W. 1879. Report on the Lands of the Arid Region of the United States. U.S. Department of Geographical and Geological Survey of the Rocky Mountain Region. 2nd ed. Washington, D.C.: U.S. Government Printing Office.

Rabinovitch-Vin, A. 1991. Continuous human use as a tool for species richness in protected areas of Israel. Pp. 95-99 in Resident People and Parks: Social Dilemmas and Strategies in International Conservation, P.C. West and S.R. Brechin, eds. Tucson: University of Arizona Press.

Randall, A. 1987. Resource Economics: An Economic Approach to Natural Resource and Environmental Policy, 2nd ed. New York: John Wiley & Sons.

Ranney, J.W., M.C. Bruner, and J.B. Levenson. 1981. The importance of edge in the structure and dynamics of forest islands. Pp. 67-

95 in Forest Island Dynamics in Man-Dominated Landscapes, R.L. Burgess and D.M. Sharpe, eds. New York: Springer-Verlag.

Rao, K., and C. Geisler. 1990. The social consequences of protected areas development for resident populations. Society and Natural Resources 3:19-32.

Reisner, M. 1986. Cadillac Desert. New York: Penguin Books.

Reitz, W.W., Jr. 1974. Land Acquisition for a Potomac National River: The Use of Less than Fee Simple Title Acquisitions to Accomplish the Objectives of the Potomac National River Concept. Completed for the Interstate Commission on the Potomac River Basin, Washington, D.C. July.

Ripple, W.J., ed. 1987. Geographic Information Systems for Resource Management: A Compendium. American Society of Photogrammetry and Remote Sensing and American Congress on Surveying and Mapping, Falls Church, Virginia.

Rodgers, W.H., Jr. 1986. 1 Environmental Law: Air and Water, Section 2.20.

Romme, W.H. 1982. Fire and landscape diversity in subalpine forests of Yellowstone National Park. Ecol. Monogr. 52:199-221.

Roush, G.J. 1991. Coming home: Reversing the American story. The Nature Conservancy Magazine May/June:17-23.

Ruckelshaus, W.D. 1989. Toward a sustainable world. Sci. Amer. 261:166-175.

Rudzitis, G., and H.E. Johansen. 1989. Amenities, Migration, and Non-Metropolitan Regional Development. Report to the National Science Foundation, Washington, D.C.

Runkle, J.R. 1985. Disturbance regimes in temperate forests. Pp. 17-33 in The Ecology of Natural Disturbance and Patch Dynamics, S.T.A. Pickett and P.S. White, eds. Orlando: Academic Press.

Runte, A. 1990. Yosemite. Lincoln, Neb.: University of Nebraska Press.

Sakolski, A.M. 1957. Land Tenure and Land Taxation in America. New York: R. Schalenbach Foundation.

Salazar, D.J., and R.G. Lee. 1990. Natural Resource Policy Analysis and Rational Choice Theory: A Strategy for Empirical Research. Natural Resources J. 30:283-300.

Salazar, D.J., and A.K. Lenard. 1992. Land Conservation and the Nature of Goods. Paper presented at the 4th North American Symposium on Society and Resource Management. Madison, Wisc. Sampson, 1980.

Sandenburgh, R., C. Taylor, and J.S. Hoffman. 1987. Rising carbon dioxide, climate change, and forest management: An overview. Pp. 113-121 in The Greenhouse Effect, Climate Change, and U. S. Forests, W.E. Shands and J.S. Hoffman, eds. The Conservation Foundation, Washington, D.C.

Sawhill, J.C. 1991a. The last great places. The Nature Conservancy Magazine, May/June:6-15.

Sawhill, J.C. 1991b. From the President. The Nature Conservancy Magazine, May/June:3.

Sax, J.L. 1970. The public trust doctrine in natural resource law: Effective judicial intervention. 68 Mich. L. Rev. 471-.

Sax, J.L. 1991. Ecosystems and property rights in Greater Yellowstone: The legal system in transition. Pp. 77-85 in The Greater Yellowstone Ecosystem: Redefining America's Wilderness Heritage, R.B. Keiter and M.S. Boyce, eds. New Haven, Ct.: Yale University Press.

Scepan, J., F. Davis, and L.L. Blum. 1987. A geographic information system for managing California condor habitat. Pp. 476-486 In Proceedings of GIS'87, 2nd Annual International Conference, Exhibits, and Workshops on Geographic Information Systems, San Francisco, Calif.

Scott, J. M. and B. Csuti. 1992. GAP Analysis: Protecting Biodiversity Using Geographic Information Systems. Idaho Cooperative Fish and Wildlife Research Unit, College of Forestry, University of Idaho, Moscow, Idaho, 83843. Unpublished memo.

Scott, J.M., B. Csuti, J.D. Jacobi, and J.E. Estes. 1987. Species richness: A geographic approach to protecting future biological diversity. BioScience 37:782-788.

Scott, J.M., B. Csuti, K. Smith, J.E. Estes, and S. Caicco. 1988. Beyond endangered species: An integrated conservation strategy for the preservation of biological diversity. Endangered Species UPDATE 5:43-48.

Seattle Times/Seattle Post-Intelligencer. 1991. Agency to announce purchase agreement to buy 845-acre ranch for $5.6 million. Nov. 17:B-1.

Short, B.C. 1989. Ronald Reagan and the Public Lands: America's

Conservation Debate, 1979-1984. College Station, Texas: Texas A&M University Press.
Simberloff, D., and J. Cox. 1987. Consequences and costs of conservation corridors. Conservation Biology 1:63-71.
Soderstrom, E.J. 1981. Social Impact Assessment: Experimental Methods and Approaches. New York: Praeger.
Solomon, A.M. 1986. Transient response of forests to $CO_2$-induced climate change: Simulation modeling experiments in eastern North America. Oecologia (Berl.) 68:567-579.
Solomon, A.M., and T. Webb, III. 1985. Computer-aided reconstruction of late Quaternary landscape dynamics. Ann. Rev. Ecol. Syst. 16:63-84.
Soulé, M.E. 1985. What is Conservation Biology? BioScience 35: 727-734.
Soulé, M.E., D. Boulger, A. Alberts, R. Sauvajot, J. Wright, M. Sorice, and S. Hill. 1988. Reconstructed dynamics of rapid extinctions of chapparal-requiring birds in urban habitat islands. Conservation Biology 2:75-92.
Sprugel, D.G. 1985. Natural disturbance and ecosystem energetics. Pp. 335-352 in The Ecology of Natural Disturbance and Patch Dynamics, S.T.A. Pickett and P.S. White, eds. Orlando: Academic Press.
Steen, H.K. 1991. The U.S. Forest Service: A History. Seattle: University of Washington Press.
Stenbeck, J.M., C.B. Travlos, R.H. Barrett, and R.G. Congalton. 1987. Application of remotely sensed digital data and a GIS in evaluating deer habitat suitability on the Tehama deer winter range. Pp. 440-445 in Proceedings of GIS'87, 2nd Annual International Conference, Exhibits, and Workshops on Geographic Information Systems, San Francisco, Calif.
Stevens, W.W. 1992. Novel Strategy Puts People At Heart of Texas Preserve. New York Times (March 31):C1, C8.
Stone, D.A., ed. 1988. Policy Paradox and Political Reason. Glenview, Ill.: Scott, Foresman.
Stolzenburg, W. 1991. The fragment connection. The Nature Conservancy Magazine. July/August:19-25.
Sullivan, A., and M. Shaffer. 1975. Biogeography of the megazoo. Science 189:13-17.

# REFERENCES

Taylor, C.N., and C.H. Bryan. 1990. A New Zealand issues-oriented approach to social impact assessment. Pp. 37-54 in Methods for Social Impact Assessment in Developing Countries, K. Finsterbusch, J. Ingersoll, and L. Llewellyn, eds. Boulder, Colo.: Westview Press.

Thomas, C.E. 1985. The Cape Cod National Seashore: A case study of federal administrative control over traditionally local land use decisions. 12 Boston College Envir. Affairs Law Revue 225-.

TNC (The Nature Conservancy). 1987. Preserve Selection and Design Manual. The Nature Conservancy, Arlington, Va.

Tosta, N., and L. Davis. 1987. Utilizing a geographic information system for statewide resource assessment: The California case. Pp. 147-154 in Proceedings of GIS'87, 2nd Annual International Conference, Exhibits, and Workshops on Geographic Information Systems, San Francisco, Calif.

Tripp, J.T.B., and D.J. Dudek. 1989. Institutional guidelines for designing successful transferable rights programs. 6 Yale Journal on Regulation. 369, 372 & n. 7.

Trust for Public Land. 1991. Letter from Ralph W. Benson, Executive Vice President to Harriet Burgess. January 18, 1991.

Tug Hill Commission. 1991. Working Lands: The Tug Hill Planning Commission. Watertown, New York.

Turner, M.G., ed. 1987. Landscape Heterogeneity and Disturbance. Ecological Studies 64. New York: Springer-Verlag.

Turner, M.G. 1989. Landscape ecology: The effect of pattern on process. Annu. Rev. Ecol. Syst. 20:171-197.

Turner, M.G. 1990. Landscape changes in nine rural counties in Georgia, USA. Photogrammetric Engineering and Remote Sensing 56: 379-386.

Turner, M.G., and R.H. Gardner, eds. 1991. Quantitative Methods in Landscape Ecology: The Analysis and Interpretation of Landscape Heterogeneity. New York: Springer-Verlag.

Turner, M.G., V.H. Dale and R.H. Gardner. 1989a. Predicting across scales: Theory development and testing. Landscape Ecol. 3:245-252.

Turner, M.G., R.V. O'Neill, R.H. Gardner, and B.T. Milne. 1989b. Effects of changing spatial scale on the analysis of landscape pattern. Landscape Ecol. 3:153-162.

Udvardy, M.D.F. 1984. A biogeographical classification system for terrestrial environments. Pp. 34-38 in National Parks: Conservation and Development, J.A. McNeely and K.R. Miller, eds. Washington, D.C.: Smithsonian Institution.

Urban, D.L., R.V. O'Neill, and H.H. Shugart, Jr. 1987. Landscape ecology. BioScience 37:119-127.

U.S. Congress, House of Representatives, 88th Cong., 1st Sess. at 12. 1963. H.R. Rep. 88-900.

U.S. Congress, Senate, 88th Cong., 2nd Sess. at 7. 1964. S. Rep. 88-1364.

USFS (U.S. Forest Service, Department of Agriculture). 1986. Land and Resource Management Plan, Mark Twain National Forest.

USFS (U.S. Forest Service, Department of Agriculture). 1987. Forest Plan, Kootenai National Forest, Libby, Montana.

USFS (U.S. Forest Service, Department of Agriculture). 1989. Forest Plan, Green Mountain National Forest, Rutland, Vermont.

USFS (U.S. Forest Service, Department of Agriculture). 1990. Land and Resource Management Plan, Mt. Baker-Snoqualmie National Forest, Pacific Northwest Region, Portland, Oregon.

USFWS (U.S. Fish and Wildlife Service). 1992. Critical Habitat and Threatened Wildlife and Plants: Determination of Critical Habitat for the Northern Spotted Owl. Final Rule. Federal Register, Vol. 57, No. 10:1796-. January 15.

U.S. Public Land Law Review Commission (PLLRC). 1970. One Third of the Nation's Land. A Report to the President and the Congress by the Public Land Law Review Commission. Washington, D.C.: U.S. Government Printing Office. June.

Waddell, T.H., D.D. Oswald, and D.S. Powell. 1987. Forest Statistics of the United States, 1987. Resource Bulletin PNW-RB-168, Pacific Northwest Research Station, Forest Service, U.S. Department of Agriculture. September.

Wakely, R. 1987. GIS and Weyerhaeuser P 20 years experience. Pp. 446-457 in Proceedings of GIS'87, 2nd Annual International Conference, Exhibits, and Workshops on Geographic Information Systems, San Francisco, Calif.

Walters, C. 1986. Adaptive Management of Renewable Resources. New York: Macmillan.

Ward, J.R., and F.K. Benfield. 1989. Conservation easements: Prospects for sustainable agriculture. 8 Virginia Environmental Law

Journal 271, 273.

Wells, P.V. 1970. Historical factors controlling vegetation patterns and floristic distributions in the central plains region of North America. Pp. 211-221 in Pleistocene and Recent Environments of the Central Great Plains, W. Dort, Jr. and J.K. Jones, Jr., eds. Special Publication 3. Department of Geology, University of Kansas. Lawrence, Kansas: The University of Kansas Press.

West, P.C., and S.R. Brechin, eds. 1991. Resident Peoples and National Parks: Social Dilemmas and Strategies in International Conservation. Tucson: University of Arizona Press.

Wheatley, C.F., Jr. 1970. Study of Land Acquisitions and Exchanges Related to Retention and Disposition of Federal Public Lands. Report prepared for the U.S. Public Land Law Review Commission. NTIS PB194448. Springfield, Va.: National Technical Information Service. September.

Whitcomb, C.E. 1987. Establishment and Maintenance of Landscape Plants. Stillwater, Okla: Lacebark Publications.

White, P.S. 1979. Pattern, process, and natural disturbance in vegetation. Bot. Rev. 45:229-299.

White, P.S., and S.P. Bratton. 1980. After preservation: Philosophical and practical problems of change. Biol. Conserv. 18:241-255.

White, P.S., and W. Cronon. 1988. Ecological change and Indian-White relations. Pp. 417-429 in History of Indian-White Relations, W.E. Washburn, ed. in Handbook of North American Indians, Vol. 4., W.C. Sturtevant, ed. Washington, D.C.: Smithsonian Institution.

White Earth Land Recovery Project. 1990. Land Use Needs Assessment for White Earth Reservation. Prepared for the White Earth Land Recovery Project, White Earth, Minnesota.

Whitney, G.T., and W.J. Somerlot. 1985. A case study of woodland continuity and change in the American midwest. Biol. Conserv. 31:265-287.

Williams, C. 1982. The Park Rebellion: Charles Cushman, James Watt, and the Attack on the National Parks. Not Man Apart. June. (A Friends of the Earth Reprint, San Francisco, Calif.)

Wilson, and Winkler. 1971. The response of state legislation to historic preservation. 36 Law and Contemporary Problems 329-.

Wolf, C.P. 1981. Land in America. New York: Pantheon.

Wolf, C.P. 1983. Social impact assessment: A methodological overview. Pp. 25-33 in Social Impact Assessment Methods, K. Finsterbusch, L.G. Llewellyn, and C.P. Wolf, eds. Beverly Hills, Calif.: Sage Publications.

Wooten, H.H. 1965. The Land Utilization Program, 1934-1964. Agricultural Economic Report No. 85. Economic Research Service, U.S. Department of Agriculture, Washington, D.C.

Worster, D. 1985. Nature's Economy: A History of Ecological Ideas. Cambridge: Cambridge University Press.

# Appendix A

## Presenters and Discussants

Keith Argow, Land Conservation Fund, Vienna, Va.
Philip Bayles, USDA Forest Service, Washington, D.C.
David Beaver, DOI, Bureau of Land Management, Washington, D.C.
Dean Bibles, DOI, Bureau of Land Management, Portland, Ore.
Chip Collins, National Fish and Wildlife Foundation, Washington, D.C.
Gary Cooper, USDA Forest Service, Portland, Ore.
Dale Crane, National Parks and Conservation Association, Des Moines, Iowa
Gina DeFerrari, U.S. House of Representatives Subcommittee on Fisheries, Wildlife, Conservation and the Environment, Washington, D.C.
Joseph Doddridge, DOI, Fish and Wildlife and Parks, Washington, D.C.
Myron Ebell, National Inholders Association, Washington, D.C.
Charles Estes, U.S. Senate Committee on Appropriations, Washington, D.C.
David Ford, National Forest Products Association, Washington, D.C.
Ronald Fowler, U.S. Fish and Wildlife Service, Washington, D.C.
David Gibbons, OMB, Natural Resources, Energy and Science, Washington, D.C.
Richard Healy, U.S. House of Representatives Subcommittee on National Parks and Public Lands Subcomittee, Washington, D.C.
John Heissenbuttel, American Forest Council, Washington, D.C.
Don Helman, The Wilderness Society, Washington, D.C.

Eric Hertfelder, National Conference of State Historic Preservation Officers, Washington, D.C.
Steven Hodapp, U.S. House of Representatives Subcommittee on National Parks and Public Lands, Washington, D.C.
James Howe, *Land Letter*, Arlington, Va.
Charles Jordan, Portland Department of Parks and Recreation, Portland, Ore.
Philip Kiko, DOI, Policy, Management and Budget, Washington, D.C.
Meredith Kimbro, DOI, Fish and Wildlife and Parks, Washington, D.C.
William Kriz, DOI, National Park Service, Washington, D.C.
Donald Knowles, DOI, Office of the Secretary, Washington, D.C.
Robert Lamb, DOI, Office of Budget, Washington, D.C.
Ronald Marcoux, Rocky Mountain Elk Foundation, Missoula, Mont.
Howard Miller, DOI, National Park Service, Washington, D.C.
Richard Moore, USFWS, Division of Realty, Portland, Ore.
Clyde Schnack, USFWS, Division of Realty, Washington, D.C.
Michael Scott, USFWS, Cooperative Unit, Moscow, Ida.
David Simon, National Park Conservation Association, Washington, D.C.
Neal Sigmon, U.S. House of Representatives Subcommittee on Interior and Related Agencies, Washington, D.C.
Donald Simpson, DOI, Bureau of Land Management, Washington, D.C.
Gordon Small, USDA Forest Service, Washington, D.C.
Robert Smith, The Cato Institute, Washington, D.C.
John Smolko, Ducks Unlimited, Long Grove, Ill.
Dennis Stolte, American Farm Bureau Federation, Washington, D.C.
Barry Tindall, National Recreation and Parks Association, Arlington, Va.
Charles Williams, Columbia Gorge Coalition, Mosier, Ore.
Joseph Wrabek, National Inholders Association/Multiple Use Land Alliance, Battle Ground, Wash.
Vincent Hecker, DOI, Bureau of Land Management, Washington, D.C.

# Appendix B

## Procedure for Compiling Federal Land Acquisition Priority List

I.  DOI and USDA determine which proposed acquisitions meet minimum criteria.

1.  The property is (a) within the boundaries of an existing Federal conservation/recreation unit, if such boundaries are set by statute; or (b) contiguous with property now comprising a Federal conservation/recreation unit, if the unit's boundaries are administratively determined; or (c) the initial "building block" of a newly authorized Federal conservation/recreation unit.

2.  The property presents no known health/safety/liability problems (e.g., hazardous waste contamination, unsafe structures).

3.  There is no indication of opposition from current owners(s) to Federal acquisition of property (condemnations may be necessary in rare instances).

4.  The cost of infrastructure necessary to make the property accessible, safe, and usable by the general public does not exceed ten percent of the estimated purchase price.

II. DOI and USDA score proposed acquisitions that meet minimum criteria.

Each potential acquisition is scored by summing points it receives from meeting one or more of the following "ranking criteria." The indicated number of points is awarded if the proposed acquisition:

1. a. Prevents *imminent* (within 2-3 years) property development that is determined by the regional or State director to be incompatible with the affected unit's authorized purpose(s). *50 points*
   b. Prevents *short-to-medium* term (within 4-8 years) property development that is determined by the Secretary to be incompatible with the affected unit's authorized purpose(s). *25 points*

2. a. Provides multiple recreation opportunities (seven or more of the activities listed on Attachment A) and is within a county with a population of one million or more. *80 points*
   b. Provides multiple recreation opportunities (seven or more of the activities listed on Attachment A) within 100 miles of a Standard Metropolitan Statistical Area (SMSA) *50 points*
   c. Provides multiple recreation opportunities (seven or more of the activities listed on Attachment A) between 100 and 250 miles from a Standard Metropolitan Statistical Area (SMSA) *35 points*
   d. Provides limited recreation opportunities (one to six of the activities listed on Attachment A) within 100 miles of a SMSA. *35 points*
   e. Provides limited recreation opportunities (one to six of the activities listed in Attachment A) between 100 and 250 miles from a SMSA. *20 points*

3. a. Preserves habitat of endangered species. *40 points*
   b. Preserves habitat of threatened species. *30 points*
   c. Preserves a recognized type of ecological community, for the purpose of promoting natural diversity. *20 points*

4. Preserves a nationally significant natural or cultural feature of a type not now represented in any Federal conservation/recreation unit. *40 points*

5. a. The principal benefit to be derived from the acquisition is its wetlands characteristics as defined in the Emergency Wetlands Act of 1986. *80 points*
b. The property contains a wetland or riparian area that is relatively scarce or unique. *60 points*
c. The property contains a wetland or riparian area that while not scarce or unique nevertheless provides substantial public benefits. *40 points*

6. a. Includes existing infrastructure required to make property accessible to and usable by the general public *and* by elderly/handicapped citizens. *40 points*
b. Includes existing infrastructure required to make property accessible to an usable by the general public, *but not* by elderly/handicapped citizens. *20 points*

7. Expands a unit with a record of visitor-day growth exceeding five percent per year in at least three of the five prior years. *20 points*

8. Improves manageability and efficiency of a unit. *20 points*

9. Results in Federal savings in acquisition costs through the use of land exchanges, donations and other alternatives to the direct purchase of a property at full value. Add *five points* for each estimated 20 percent savings in Federal acquisition costs up to a maximum of 25 points. *5-25 points*

10. Involves Federal acquisition of less than full fee title to the property (e.g., purchases of scenic or conservation easements). *10 points*

11. Involves significant non-Federal partnership. For each non-Federal partner (State, local, or private) contributing significant resources (i.e., at least 25 percent of acquisition, development, or management dollars), add *5 points*, up to a maximum of 15 points. *5-15 points*

12. Provides a Federal Land Management Agency with an opportunity judged by the appropriate Assistant Secretary to be necessary to substantially further the goals of a Presidential, Departmental, or Bureau MBO and to be essential to the fulfillment of the Agency's mission. (Each Agency will rank their 20 highest priority projects which implement their MBO and mission in descending order. The first priority project will be awarded 150 points, the second 142.5, the third 135 and son on). *7.5-150 points*

Ranking criteria listed under a single number (e.g., "1.a.," "1.b.") are mutually exclusive; points may not be awarded for more than one. A proposed acquisition may score 40 points for meeting criteria 3.a., but cannot score 70 points for meeting both criteria 3.1. and 3.b.

Each property proposed for acquisition is normally scored separately. However, if several related properties are proposed for purchase as a group to optimize use of funds, the group may be assigned a composite score (e.g., the criteria may be applied and a score assigned to the Forest Service Lake Tahoe project or the Park Service Santa Monica Mountains project as a whole.).

III. DOI and USDA tentatively rank acquisition proposals.

The Departments jointly prepare a unified list of acquisition proposals, ranked in descending order of points scored. A cut-off point is determined by summing costs associated with the ranked acquisitions until the cumulative cost reaches the land acquisition budget limit ($250 million, less administrative and emergency acquisition costs).

IV. Review by Land Acquisition Working Group.

The Land Acquisition Working Group, including representatives of the Assistant Secretary of the Interior for Fish, Wildlife and Parks, the Assistant Secretary for the Interior for Land and Minerals Management, and the Assistant to the Secretary of Agriculture for Natural Resources and Environment, reviews and modifies the tentative ranking of land acquisition proposals to reflect (1) corrections of identified scoring errors, (2) proposed exceptions to the minimum criteria, and (3) subjective factors not taken into account in the scoring process.

Examples of subjective factors include, but are not limited to: the role of a given acquisition in a coordinated Federal/State/local effort to preserve recreation lands; the possible effect of an acquisition on State, local, or private efforts to offer competing recreation opportunities; the prospect that a private conservation group may desire to purchase the property.

For each proposed acquisition added to the list on a basis other than points scored, the Working Group prepares a written justification explaining why the acquisition has been afforded higher priority.

## Attachment A

### Recreation Activities

Hiking
Camping
Picnicking
Fishing
Hunting
Swimming
Boating/Canoeing/Rafting
Auto Touring
Off-Road Vehicle Use
Skiing/Ice Skating
Snowmobiling
Bicycling
Horseback Riding
Observing Wildlife

# Appendix C

## National Surveys Relevant to Public Land Use, Protection, and Purchase

1. Now let me ask you about a few specific federal agencies (here, NPS or National Park Service)—Is your opinion of them highly favorable, or moderately favorable, or not too favorable?

|  | Percentage |
|---|---|
| Highly favorable | 37 |
| Moderately favorable | 43 |
| Not too favorable | 8 |
| Unfavorable | 3 |
| Don't know | 9 |

*Organization conducting survey: Roper Organization*
*Source: Roper Report 87-8*
*Interview method: personal (n = 1,967 U.S. adults)*

2. I am going to read you a list of current national issues. I'd like you to tell me how concerned you are about each issue, using a scale of 1 to 5, which 1 is not too concerned and 5 is extremely concerned. . . . "Establishing wilderness areas throughout the country."

|  | Percentage |
|---|---|
| Extremely concerned (pt. 5) | 34 |
| Not extremely concerned (pt. 1-4) | 66 |

*Organization conducting survey: Opinion Research Corporation*
*Source: Public Opinion Index*
*Interview method: telephone (n = 1,011 U.S. adults)*

3. Within the last five years, have you visited a federally owned and managed park area such as a national park, forest or monument?

|  | Percentage |
|---|---|
| Yes | 62 |
| No | 36 |
| Don't know | 2 |

*Organization conducting survey: Market Opinion Research*
*Source: Participation in Outdoor Recreation, President's Commission on Americans Outdoors*
*Interview method: telephone (n = 2,000 U.S. adults)*

4. Do you favor or oppose the following proposals that are now being considered by the federal government? . . . Spending money to improve the condition of the national parks rather than expanding the national-park system.

|  | Percentage |
|---|---|
| Favor | 73 |
| Oppose | 20 |
| Don't know | 7 |

*Organization conducting survey: Gallup Organization*
*Source: Gallup/Newsweek (1981)*
*Interview method: telephone (n = 745 U.S. adults)*

5. Do you favor or oppose the following proposals that are now being considered by the federal government . . . "Increasing oil exploration and other commercial uses of federal lands (including national parks)?"

|  | Percentage |
|---|---|
| Favor | 76 |
| Oppose | 19 |
| Don't know | 5 |

*Organization conducting survey: Gallup Organization*
*Source: Gallup/Newsweek (1981)*
*Interview method: telephone (n = 745 U.S. adults)*

6. Are you in favor of setting aside more public land for conservation purposes such as parks, wildlife refuges, bird sanctuaries, and so forth, or not?

|  | Percentage |
|---|---|
| Yes | 75 |
| No | 19 |
| Don't know/no answer | 6 |

*Organization conducting survey: Gallup organization*
*Sponsor: National Wildlife Federation (1969)*
*Interview method: personal (n = 1,503 U.S. adults)*

7. I am going to read you a number of new proposals that have been suggested for using government revenues in order to reduce the federal budget deficit. For each one, indicate whether you favor or oppose it. . . . Selling the National Wilderness areas?

|  | Percentage |
|---|---|
| Favor | 5 |
| Oppose | 89 |
| Don't know | 6 |

*Organization conducting survey: Gallup Organization*
*Sponsor: Times Mirror (People, the Press & Politics, 1988)*
*Interview method: personal (n = 3,021 U.S. adults)*

8. How about their [environmental groups] efforts to protect and expand the national wilderness areas. Do you feel that you and your family benefit?

|  | Percentage |
|---|---|
| Great deal | 30 |
| Fair amount | 34 |
| Only a little | 23 |
| Not at all | 11 |
| Don't know/no answer | 3 |

*Organization conducting survey: Bureau of Social Science Research*
*Sponsor: Resources for the Future (National Environmental Survey, 1978)*
*Interview method: telephone (n = 1,076 U.S. adults)*

9. Do you think the U.S. government should sell some national forest land to private organizations or not, or is that something you don't have an opinion on?

|  | Percentage |
|---|---|
| Yes, it should sell | 11 |
| No, it should not sell | 58 |
| No opinion | 31 |

*Organization conducting survey: ABC News/Washington Post*
*Source: ABC News/Washington Post (1983)*
*Interview method: telephone (n = 1,516 U.S. adults)*

# Appendix D

## The Nature Conservancy: Aquisition Priorities and Preserve Selection and Design

### ACQUISITION PRIORITIES

In 1917, the Ecological Society of America (ESA) formed the Committee for the Preservation of Natural Conditions for Ecological Study, chaired by Victor Shelford. For 26 years, that committee tried to locate and preserve natural areas for scientific research, and it actively supported conservation organizations. Some of the membership objected to ESA's active support policy, and in 1946, Shelford formed a private independent organization, The Ecologists' Union, which eventually published an inventory of 691 nature sanctuaries. In 1950, The Ecologists' Union was renamed The Nature Conservancy (TNC). McIntosh (1985) stated "it has proven markedly successful in securing natural areas using a revolving fund and arranging for their protection, either by transferring responsibility to a stable institution in state or federal government or by managing it themselves."

### Goals

TNC states that its primary objective is to conserve biological and ecological diversity. To this end, TNC lists, classifies, characterizes, and inventories "the enormous diversity and complexity of our biota, ecosystems, and landscapes." The National Heritage Program Operation Manual (NHPOM) states, "many earlier conservation inventories suffered from objectives that were unclear or overly general. By lumping together information on biology, scenery, recreation values, exploitable resource potential, property ownership, etc. the results were muddled."

To accomplish TNC's objective, natural heritage inventory programs were established. Those programs are defined as

> a permanent and dynamic atlas and data bank on the existence, identity, characteristics, numbers, condition, status, location and distribution of the elements of natural biological and ecological diversity, of the individual occurrences of these elements in the landscape, of existing preserves, of outstanding sites of potential preserves, of associate land ownerships, and of sources of additional information and documentation.

The programs (which are updated continually) collect, manage, and use biological, ecological, and related information in cooperation with various state agencies. With the formation of the first state natural heritage program in South Carolina in 1974, a systematic and workable approach was developed by incorporating element occurrence[1] (EO) and ranking as a core to classify natural areas. Heritage programs (also known as conservation data centers) have been established in all 50 states, 13 Latin American countries, 4 Canadian provinces, the Caribbean, and the South Pacific. During the past 40 years, TNC has acquired 5.5 million acres in the United States and Canada and 15 million acres in Latin America (Sawhill, 1991b).

## Priorities

Because it is impossible to inventory all biological diversity, heritage programs tend to focus on the rarest, most endangered, and most vulnerable species (including infraspecific taxa) and communities with a special emphasis on vertebrates and vascular plants. However, micro-organisms, nonvascular plants, and invertebrates do receive attention if specialists think they are imperiled. TNC selects "the last of the least, and the best of the rest." Attention is devoted to a comprehensive inventory

---

[1] An element occurrence is any type of biological or ecological entity, e.g., species or community, in a geographic area.

of communities and ecosystems within each state "at a practical level of discrimination."

Priority ordinal ranking is based upon "elements of natural diversity," which NHPOM defines as "the basic units of the classification system and the targets of the heritage inventory. These units are natural entities which . . . represent the full array of natural diversity for the state or region covered." Sites are ranked for the rarity of the elements and the quality of the element occurrence as well as other values, uses, or benefits a site might have in addition to its potential contribution to biological diversity conservation. Those values include ecological service functions, such as aquifer recharge and erosion prevention, and other benefits, such as recreation, aesthetic enjoyment, and historic and archeological significance. This helps to identify potential partnerships and increase the feasibility of implementation (R. Jenkins, pers. comm., TNC, 1991). The goal is for all species to receive a "global" conservation rank. All North American vertebrates and 90% of the North American vascular plants have been assigned ranks, as have many other taxa.

## Operations

Heritage programs are established within states when a state agency or other institution expresses a desire to cooperate and agrees to take over full support after start up. In the absence of expressed cooperation, TNC can begin operations entirely on private funding. The staff is hired by TNC in consultation with the cooperating agency and typically consists of a botanist, a zoologist, a community ecologist, and an information manager.

Operations are detailed in NHPOM, which ensures that all heritage programs have standard data collections and information storage. The manual is a technical one; specific information to collect usually is left to the discretion of the staff in each state and is, in part, dependent on the information base already accrued in each state.

## Why Heritage Programs Work

NHPOM lists the following principles as accounting for the success of heritage programs.

**Focused goal:** The clearly stated goal is to conserve natural diversity and to collect information on selected subsets of "all the possible landscape attributes." That restricted focus and standardized data management have facilitated the accumulation of a valuable and highly usable data base.

**Common units of comparison:** The heritage program developed the concept of elements of diversity that are defined as species, community types, or other special features. The standard nomenclature is "element, element occurrence, site (land unit of preserve design), tract (land ownership parcel), managed area (for a preserve or semiprotected area), and source of information (publication, person, agency, file, etc.).

**Balanced information system:** The information system consists of computerized and manual files. Computer records are used for sorting, report generation, and to answer specific questions. The result is a permanent but continually growing data bank.

**Factual information:** The data base does not contain interpretations, conclusions, or weighting values. The system contains "what the actual elements are, what their characteristics are, exactly where on the landscape their occurrences can be found, how their locations and geographic extent relate to ownership tracts and existing preserves, etc." Thus, an accumulative and factual data base develops that can be used by a wide range of users for a variety of purposes.

**Multi-institutional cooperation:** State agencies typically conduct the heritage inventories. TNC provides a standard methodology, initially trains the staff, offers technical support, and coordinates data exchange and multistate collaboration. In addition, TNC often provides private funds for initial operations. As a result, heritage programs become permanent within state governments but are open for cooperation with other agencies and organizations. And the standard methodology provides a uniform and highly integrated data base.

**Operational continuity:** The heritage date base is a permanent inventory that grows and evolves with time. Of importance is the establishment of an institutional memory that "becomes increasingly accurate, complete, and useful." The inventory includes information on what works and what doesn't work.

**Practical geographic scale:** The location of heritage programs within states has achieved a geographic scale that is manageable and comprehensive.

**Successive approximation:** The heritage program and its data base change over time as methods become refined and new data accumulate.

**Data ranking:** A system of element priority ranks was designed to focus data gathering on the rarer elements; inventories are not bogged down with limitless information.

**Standardization:** The heritage program "exhibits an unprecedentedly high degree of standardization throughout. Absolute uniformity is maintained," which is crucial for data exchange, efficient research, system evolution, and data retrieval for users."

**Central support, data bases, and networking:** The science division trains new staff, develops new procedures, compiles suggestions from field offices, raises funds, maintains operating manuals, monitors individual programs, facilitates interaction among programs, encourages cooperation with other agencies and institutions, and maintains a central data base.

**Cooperation of the local TNC office:** TNC's state field offices are required to support the heritage programs in a variety of ways that are beneficial for all parties.

**Utility:** Heritage data systems originally were designed to foster biological conservation but have proven to be useful in other areas, such as research, education, management, environmental impact review, and development planning.

**Objective neutrality:** The heritage program tries to maintain a stance of objectivity to protect the integrity and credibility of the data base.

## The Classification System

Faced with the enormous task of inventorying and preserving biological diversity, TNC has set up a system of filters. The coarse filter is designed to include most of the species present without dealing with individual species. This is accomplished by classifying community types. With the preservation of most community types in a region, TNC estimates that 85-90% of the biological diversity of a state will be preserved. However, some rarer species might be found only in a few communities. Thus, a fine filter is imposed that lists individual species and where they occur. NHPOM discusses how to decide which species should be fine filtered, and instructions also discuss potential problems with subspecies and hybrids.

Once listed, pertinent information on a species' status is entered into the data base for tracking. Manual files, including maps, are created. Heritage staff are advised to concentrate initially on managed land in compiling element lists. Managed land usually is in public or institutional ownership with a professional manager or managing agency. This information is important to heritage programs because: 1) a large proportion of the biological diversity of a state often occurs on managed land, so conservation efforts will not be wasted on an element already protected to some degree, 2) it is often easier to increase protection on managed lands than on unprotected lands, and 3) managed areas often have rich data bases that may be used by the heritage program.

## Priorities for Acquisition

After a preliminary inventory information base is established within a state, elements are ranked based upon data collected on the frequency of occurrence of an element. The ranking is done at a global (G), national (N), and state (S) level. At each of these levels, an element is ranked from 1-5, with 1 being most critical and 5 being least critical; e.g., a G1 ranking means "critically imperiled globally because of extreme rarity (5 or fewer occurrences or very few remaining individuals or acres) or because of some factors making it especially vulnerable to extinction." Subspecies may be ranked by attached a T and 1-5 to the global ranking.

When the ranking has been decided for an element, measures are recommended that should be taken to conserve the element, including inventory, research, and stewardship.

Noss (1987) criticized TNC's system for not considering the relationships of community types within real landscapes. Noss recommended that TNC's coarse filter be expanded so that 1) disturbance and regeneration patterns are included in the evaluation, 2) landscape mosaics are addressed, and 3) surrounding habitat and corridors are examined in conjunction with the EO. TNC has recognized this problem and is following these recommendations.

TNC recently started a biological reserve initiative called the "Last Great Places": 12 sites (8 in the U.S. and 12 in Canada) ranging in size from 12,000 to 1,000,000 acres have been targeted for preservation. Each area has been judged as an ecologically salvageable landscape and a functioning, but endangered, ecosystem that contains rare species (Sawhill, 1991a). All 12 sites have a core natural zone of critical habitat and a surrounding buffer zone in which land-use practices could affect the core. According to the TNC president, the initiative fully recognizes the rights of residents in these areas and will attempt to design stewardship practices that are compatible with conservation and human interests (Sawhill, 1991a). This initiative will be science driven, with the twofold objective of conservation and sustained yield. This requires emphasis on the integration of humans into the conservation equation through sustainable development schemes and cooperative management strategies for multiuse landscapes. In the future, project selection process probably will favor sites that are or can be included in landscape complexes over isolated sites of dubious long-term viability (R. Jenkins, pers. comm., TNC, 1991).

## PRESERVE SELECTION AND DESIGN

### Objective

The *Preserve Selection and Design Manual* (PS&D) (TNC, 1987) is to help select the highest priority sites for protection of EOs based upon information supplied by the heritage program data base. This manual deals primarily with the administrative procedures of designing reserves and "does not deal at length with the scientific aspects of preserve design."

The protection activities rely on three information sources provided by heritage programs:

1) The *Natural Diversity Scorecard* lists elements in order of their

global and state ranking and scores how well they are doing in terms of protective status: well-protected, underprotected, or not protected.

2) The *Site Tracking Record* lists alphabetically the most important sites to be protected, the EOs at these sites, the ownership, and protection status by ownership tract.

3) The *Priority Site Lists* is produced by analyzing the site tracking record and listing sites in order of significance that will contribute the most to natural diversity preservation. These sites become the priorities within states for preserve design.

## The Administrative Process

The preserve design process has five phases:

1) Initiation of preserve design (site selection, budgeting, hiring designer);
2) Preparation of a preserve design package by
   a. identifying and mapping boundaries adequate to achieve the conservation objective,
   b. identifying potential threats to the site,
   c. discussing stewardship requirements,
   d. assessing ownership;
3) Analysis of package by state and regional director;
4) Assignment of protection levels
   a. voluntary by owner/manager,
   b. bequest,
   c. legally binding protection agreement,
   d. landowner-conveyed interest to conservation entity,
   e. public agency agreement to conservation designation,
   f. less-than-fee acquisition,
   g. dedication or trust investiture to conservation trust or established nature preserve system;
5) Development of protection strategy, including
   a. stewardship needs,
   b. required level of protection,
   c. cost effectiveness for desired level of protection,
   d. adequate information for TNC use.

As a result of this process, a preserve design package is constructed that is considered "the unit of proposal and approval" and should contain all the relevant information necessary for TNC to make a decision about preserving a tract.

The acquisition is approved and funded at the state level, but for projects more than $200,000 regional approval is required; for acquisitions more than $500,000, approval by the national board is necessary (Ben Pierce, pers. comm., TNC, Wyoming Field Office, 1991).

The PS&D manual contains a memorandum describing the current effort to computerize the design process. A standard form is available to design planners that prescribes information to be included.

## How To Design a Preserve

The PS&D describes how to design a preserve, but the discussion is prefaced by a caveat:

> None of these materials, however, will quite give you the answers you want or tell you how to design a preserve for particular elements and there is nothing in them that can be distilled into a set of generally recognized and incontrovertible rules. . . . Some central questions are what the minimum viable size a reserve must be to retain the greater part of its biota over the long term and what the minimum viable population a species must have to avoid a significant loss of alleles and eventual extinction on a site.

Sites are categorized as megasites (more than 64,000 acres), macrosites (more than 3,200 acres), or standard sites (fewer than 3,200 acres). Mega- and macrosites are categorized according to the following priorities: 1) unique sites with large numbers of endemic and nearly endemic species, 2) sites containing the range or most of the range of an endemic species, 3) buffers and/or corridors to protect larger, far-ranging species, and 4) representative sites of presettlement ecosystems.

Each proposed mega- and macrosite is ranked for 1) quality/significance/uniqueness/representativeness, 2) condition, 3) viability, and 4) defensibility/manageability with a letter grade from A (excellent) to D (poor).

## Minimum Viable Size of Preserves

The PS&D states that preserves may lose species over time, particularly if they are small, but the causes of species loss are not well understood, and the probability of loss differs among taxa. Many plant species (e.g., prairie flora) have existed for a long time on small plots. Practicality is often the overriding issue: "big" may be best, but big is also expensive and frequently nonexistent because of habitat loss and fragmentation.

The PS&D uses the term "minimum viable size of preserves," although it is generally understood that minimal viable preserve size varies with the EOs targeted for preservation. A hectare may be sufficient to preserve a self-fertilizing, long-lived plant, but a half million hectares in a fragmented landscape may be insufficient to preserve a population of a large vertebrate predator. It might be possible, nonetheless, to classify different types of EOs and provide some general guidelines for optimal to minimal reserve sizes.

## Minimum Viable Population Size

TNC recognizes that small populations are more subject to extinction than large populations for genetic and environmental reasons, but states that minimal population size cannot realistically be determined by analyzing any one factor. The PS&D concludes that each preserve is an experiment in minimal population size, and reserve designers should try to prescribe boundaries and management practices that minimize extinction. TNC tries to ensure that rare EOs are preserved in enough places "that the likelihood of all going extinct at the same time is vanishingly small."

The committee believes that more precise and relevant terminology could be developed related to population and genetic factors. In addition, it might be possible to develop more general guidelines than those mentioned above drawing upon knowledge of past extinctions, evolutionary trends, and the genetic structures of populations. Even if generalizations were not possible, a list of factors could be provided that could guide more objectively the preserve designer. Available information on factors for EOs designated for preservation could be collected. Examples of factors would include genetic structure, population size, effective breeding population size, breeding structure, longevity, net reproductive rate, degree of niche specialization, stability/predictability of local climate (from long-term climate data), probability of natural catastrophes (by looking at historical records), and landscape information. Much of this information is not available for most species. But an ordinal ranking scale could be developed as done for the EO.

TNC might be reluctant to adopt such a system. If the ranking suggests that a preserve may be too small to sustain an EO for a reasonable length of time, the mission of TNC directs it to err on the conservative side: that is, even if the best scientific information suggests a low chance of success, TNC may attempt preservation, because extinction is forever.

TNC acknowledges that "there are so many factors to consider in doing preserve design that we cannot enumerate them all. In future years . . . we would like to draw up a list of these factors, plus provide a good bibliography on preserve design."

## Preserve Configuration and Justification

The PS&D states that preserve design theory suggests that preserves should be as round as possible (to minimize edge) and connected by corridors for facilitating migration. But in all practicality, TNC admits that preserve shape and juxtaposition are difficult to control because of ownership patterns and past habitat disturbance. If an EO does become extinct on a preserve, TNC will consider reintroduction from elsewhere.

This criterion show that TNC works within the constraints of the real world. Tracts available for preserves tend to be much smaller than desirable, but accumulating many unconnected tracts containing rare EOs might reduce the probability of extinction; if local extinctions occur, reintroductions are possible.

The concept of corridors is not well defined, according to a member of TNC's board of directors (Stolzenberg, 1991), and corridors are considered an inadequate substitute for suitable habitat:

> I'm a skeptic about corridors. What I'm not a skeptic about is large habitat as a requirement for viable persistence of species. The most important thing is

always to identify the remaining habitat of the most endangered species and ecosystem types. If you don't save those, the things that occupy them are simply gone (Stolzenberg (1991) quoting TNC Director of Science, Robert Jenkins.)

Simberloff and Cox (1987) point out also that corridors can provide access for parasites, predators, pests, fire, and poachers.

Noss and Harris (1986) are strong advocates of corridors, and Stolzenburg (1991) described three studies that demonstrate the efficacy of corridors for maintaining populations: Harper in Brazil found that corridors were essential for maintaining antbirds in patches of jungle, Bennett in Australia discovered that corridors provide both transportation and a conduit for gene exchange, and Merriam in Canada found that woodlots connected by wooded fencerows demonstrated a continual process of extinction and recolonization by small mammals and birds.

In practice, some TNC conservation efforts have integrally incorporated corridors. Merrill Lynch (TNC, North Carolina) designed "the granddaddy of all corridors" in a 30,000 acre reserve in Pinhook Swamp in northern Florida, which was a cooperative effort between TNC and the Forest Service. Lynch is planning a multicorridored project, 436,000 acres of the Alligator River Wildlife Refuge (in collaboration with the USFWS, TNC, and the Conservation Fund), targeted for black bear and red wolf protection (Stolzenburg, 1991). And TNC's "The Last Great Places" initiative is clearly directed at saving landscapes with core habitat surrounded by buffer zones.

## Element and Site Stewardship

The PS&D insists that proper stewardship requires highly specific information about an element and the site it occurs on. It may take several years to collect the needed natural history of an element and the possible constraints on managing for its survival. The PS&D manual, however, does not help the field staff set priorities based upon scientific information on what information should be collected. The manual does not provide a review of pertinent ecological or conservation literature from either an empirical or theoretical perspective.

This criterion of TNC acquisition strategy shows an attentiveness to practicality. PS&D briefly dismisses most theoretical issues concerning preserve design and focuses on what site-specific information is needed and what can be obtained. For example, for an endangered plant, what are pollinators and seed dispersal agents? What is soil type and habitat affinity? Are there threats from neighboring farmers through herbicide use?

This type of detective work is absolutely necessary, and obviously successful, based upon the history of TNC preservation efforts. Site-specific data from field surveys is one key to TNC success. It is important, though, that data gatherers be instructed in the empirical and theoretical literature that may cast light on how to monitor and measure the ecological health of a particular site. The PS&D manual and its successor documents may be one way to accomplish this.

# Glossary

**ANCILA:** Alaska National Interest Lands Conservation Act
**BLM:** Bureau of Land Management, Department of the Interior
**CEQ:** Council on Environmental Quality
**CRP:** Conservation Reserve Program
**DOI:** Department of the Interior
**DU:** Ducks Unlimited
**EIA:** Environmental Impact Assessment
**EO:** Element occurence, used in the acquisition criteria of The Nature Conservancy
**ESA:** Endangered Species Act
**EWRA:** Emergency Wetland Resources Act
**FLPMA:** Federal Land Policy Management Act
**GIS:** Geographic information system
**HCRS:** Heritage Conservation and Recreation Service
**Inholder:** Individual or entities holding property inside boundaries of federal holdings
**LAPS:** Land Acquisition Priority System of the Office of Management and Budget
**LWCF:** Land and Water Conservation Fund
**MBCA:** Migratory Bird Conservation Act
**NAWCA:** North American Wetlands Conservation Act
**NAWCC:** North American Wetlands Conservation Council
**NAWMP** North American Waterfowl Management Plan
**NEPA:** National Environmental Policy Act

**NFMA:** National Forest Management Act
**NFWF:** National Fish and Wildlife Foundation
**NHPA:** National Historic Preservation Act
**NPS:** National Park Service, Department of the Interior
**NWPCP:** National Wetlands Priority Conservation Plan
**OMB:** Office of Management and Budget
**ORRCC:** Outdoor Recreation Resources Review Commission
**PLLRC:** Public Land Law Review Commission
**PS&D** *Preserve Selection and Design Manual* of The Nature Conservancy
**SIA:** Social impact assessment
**TDRs:** Transferable development rights
**TNC:** The Nature Conservancy
**TPL:** Trust for Public Land
**USFS:** United States Forest Service, Department of Agriculture
**USFWS:** United States Fish and Wildlife Service, Department of the Interior